Rendezvous in Distributed Systems

Zhaoquan Gu · Yuexuan Wang
Qiang-Sheng Hua · Francis C.M. Lau

Rendezvous in Distributed Systems

Theory, Algorithms and Applications

 Springer

Zhaoquan Gu
Guangzhou University
Guangzhou
China

and

The University of Hong Kong
Hong Kong
China

Yuexuan Wang
Zhejiang University
Hangzhou
China

and

The University of Hong Kong
Hong Kong
China

Qiang-Sheng Hua
Huazhong University of Science
    and Technology
Wuhan
China

Francis C.M. Lau
The University of Hong Kong
Hong Kong
China

ISBN 978-981-10-9939-7   ISBN 978-981-10-3680-4   (eBook)
DOI 10.1007/978-981-10-3680-4

Printed on acid-free paper

This Springer imprint is published by Springer Nature
The registered company is Springer Nature Singapore Pte Ltd.
The registered company address is: 152 Beach Road, #21-01/04 Gateway East, Singapore 189721, Singapore

To my family who have supported me in the past years.

—Zhaoquan Gu

To my husband and son, Giulio and Francesco.

—Yuexuan Wang

To my family, especially to my son, Hua Hua.

—Qiang-Sheng Hua

To all readers who are interested in math and computer science.

Preface

A distributed system is a collection of autonomous units able to make decisions locally. Through cooperation, these distributed units can solve global computational problems together. This book is dedicated to one of the most fundamental processes in operating a distributed system, referred to as the *rendezvous* process. Rendezvous takes place when the external ports belonging to two neighboring units become connected, and they can communicate and exchange information through this connection.

There are five dimensions in which to define an instance of rendezvous. First of all, the autonomous units can run an algorithm that is *symmetric* or one that is *asymmetric*. In some distributed systems, the autonomous units all play different roles and so they may run different algorithms, which are also called asymmetric algorithms, while in other systems, all units are of the same type and run the same symmetric algorithm. Second, the distributed units may or may not start at the same time, which correspond to the *synchronous* or *asynchronous* scenario, respectively. Third, some of the external ports in a unit may be occupied by services unrelated to rendezvous, and so different units may have different sets of available ports. It is the *symmetric port setting* if all units have the same set of available ports, otherwise *asymmetric port setting*. Fourth, the distributed units are *anonymous* if they appear to be indistinguishable, or *non-anonymous* if they can be distinguished by their labels or identifiers. Labeling applies to external ports also. In some distributed systems, the external ports have the same labels across all units, which is referred to as *non-oblivious port labeling*. These global labels of the ports simplify the rendezvous problem as one can easily capitalize on these labels in designing the rendezvous solution. Alternatively, *oblivious port labeling* assumes there is no global labeling rule, and the units label their external ports locally. Obviously, rendezvous is harder to achieve when using the latter labeling which however is more practical. In this book, we present rendezvous algorithms for all the combinations of these five dimensions.

This book is divided into five parts, which are further broken into 20 chapters. We start with an introduction of distributed rendezvous theory, which includes distributed system preliminaries, distributed computing, and rendezvous theory.

Next, we present different kinds of rendezvous algorithms for the blind rendezvous problem in Part II, where the autonomous units' ports have the same labels, i.e., non-oblivious port labeling. Then, in Part III, we introduce the oblivious blind rendezvous problem where the ports are labeled locally by the units and we present distributed rendezvous algorithms for a range of rendezvous settings. In Part IV, we introduce several rendezvous applications and discuss the method of extending the rendezvous algorithms for distributed systems to these applications. Finally, we summarize the rendezvous results and mention some future work in Part V.

This book can be treated as a handbook of solutions to the rendezvous problem in distributed systems. Rendezvous as a fundamental process underpins the construction of many important functions in distributed systems and networks. Other than theories and algorithms, this book also covers applications in which rendezvous has a valuable role to play. These applications are just a small sample of many potential applications that can benefit from an efficient rendezvous process. This book offers in particular an in-depth treatment of the blind rendezvous and oblivious blind rendezvous problems and their solutions. Rendezvous should be of interest to readers from other research fields such as robotics, wireless sensor networks, and game theory as the need for rendezvous arises naturally in many scenarios in these different fields.

"If I had my life to live over, I'd have fewer meetings and more rendezvous" (Robert Breault). Indeed, rendezvous is more than just a usual meeting which might not amount to anything; there is a purpose behind every rendezvous which is to enable the parties involved to establish a relationship and engage in an activity that will benefit both. Rendezvous makes things happen!

Should you have any questions or suggestions, please contact the authors via e-mail to *zqgu@hku.hk*, *amywang@zju.edu.cn*, *qshua@hust.edu.cn*, or *fcmlau@cs. hku.hk*.

Beijing, China   Zhaoquan Gu
March 2017   Yuexuan Wang
                                                                                 Qiang-Sheng Hua
                                                                               Francis C.M. Lau

Acknowledgements

The authors would like to express their gratitude to all individuals who have helped us during the revision of the manuscript, in correcting mistakes and suggesting new examples. We are indebted to Editor Xiaolan Yao who provided plenty of useful information during the entire preparation. This research was supported in part by the Hong Kong Scholars Program.

Contents

Acronyms

AAG	Adversary assignment graph
AHW	Alternate hop-and-wait
Alg	Algorithm
A-QCH	Asynchronous quorum-based channel hopping
ARPANet	Advanced Research Projects Agency Network
BR	Blind rendezvous
BRP	Blind rendezvous problem
CBH	Conversion-based hopping
CCC	Common control channel
CH	Channel hopping
CHAT	Channel hopping multiple access with packet trains
CHMA	Channel hopping multiple access
CR	Cognitive radio
CRN	Cognitive radio network
CRSEQ	Channel rendezvous sequence
CTS	Clear-to-send
DC	Distributed computing
DCA	Dynamic channel allocation
DCA-PC	Dynamic channel allocation with power control
DPC	Dynamic private channel
DRDS	Disjoint relaxed difference set
DRSEQ	Deterministic rendezvous sequence
DS	Distributed system
DSA	Dynamic spectrum access
DT	Distributed theory
EJS	Enhanced Jump Stay
ESA	Expanded sequential accessing
ETTR	Expected time to rendezvous
FPA	Fixed port accessing
FTP	File transfer protocol

GOS	Generated orthogonal sequence
GS	Global sequence
HCRN	Heterogeneous cognitive radio network
HH	Heterogeneous hopping
ID	Identifier
IDH	ID hopping
IEEE	Institute of Electrical and Electronics Engineers
ISM	Industrial, Scientific, and Medical
JS	Jump stay
LAN	Local area network
LB	Lower bound
LS	Local sequence
MAC	Media Access Control
MAP	Multichannel access protocol
MC	Modular clock
McMAC	Multichannel MAC
MLS	Modified local sequence
MMC	Modified modular clock
MPM	Message passing model
MSA	Modified sequential accessing
MSH	Multistep port hopping
MTP	Moving traversing pointer
MTTR	Maximum time to rendezvous
MWN	Multichannel wireless network
ND	Neighbor Discovery
NP	Non-deterministic polynomial
NPC	Non-deterministic polynomial complete
OBR	Oblivious blind rendezvous
OSA	Oblivious sequential accessing
PSTN	Public switched telephone network
PU	Primary user
QCH	Quorum-based channel hopping
QS	Quorum system
RDS	Relaxed difference set
RS	Rendezvous setting
RSp	Rendezvous search problem
RW	Ring-walk
SA	Sequential accessing
SAA	Sequential accessing algorithm
SCH	Synchronous check and hop
SDS	Singer difference set
SMM	Shared memory model
SPA	Smallest port accessing
SRS	Stay or random selection
SSCH	Slotted seeded channel hopping

SU	Secondary user
TP	Traversing pointer
TTR	Time to rendezvous
TV	Television
TWA	Temporary wait algorithm
UHF	Ultra-high frequency
VHF	Very high frequency
w.l.o.g	Without loss of generality
WAN	Wide area network
WSN	Wireless sensor network
WWW	World Wide Web

Part I
Distributed Rendezvous Theory

Chapter 1
Distributed Systems

Abstract Distributed systems as an implementation choice have become dominant in both scientific research and practical applications. As a collection of autonomous units which can make decisions locally, distributed systems are capable to solve many complex computational problems through appropriate cooperations among the units. In this chapter, we introduce distributed systems and describe some of their properties. There are many examples of distributed systems in our daily life; we mention some of them for a better understanding of the concepts of distributed systems. In Sect. 1.1, we introduce the concept of distributed systems and some important properties of a distributed system. Then, we highlight several real-life examples of distributed system in Sects. 1.2–1.6, including local area networks, email network systems, wireless sensor networks, cognitive radio networks, and telephone networks.

1.1 What is Distributed System?

A distributed system is a distributed collection of computing units that can make decisions locally. Different from a *centralized system*, which is usually composed of many units that are attached to a single controlling unit, distributed systems make possible carrying out a global task collectively by the distributed units without any centralized control.

Originally, distributed systems refer to computer networks where the computers are physically distributed. The term is used in a much wider sense nowadays, such as wireless sensor networks where the distributedly located sensors are connected as a network, networks composed of robots that are dispatched to carry out some tasks, and ad-hoc networks of mobile phones that happen to be in the same locality. While there is no standard definition of a distributed system that would fit all situations, the following properties are commonly considered and found in a distributed system:

(1) The entities in the system can be any computational unit or node, such as a computer, a mobile phone, a robot, or even people;
(2) the entities are autonomous, that is, each entity can make decisions by itself;

© Springer Nature Singapore Pte Ltd. 2017
Z. Gu et al., *Rendezvous in Distributed Systems*,
DOI 10.1007/978-981-10-3680-4_1

(3) the entities in the system can be connected as a network, but there exists no oracle (such as a centralized unit or an overall controlling unit) informing the entities of the network's information;

(4) the entities may communicate with each other to exchange their information.

Having these properties, distributed systems can be more than just comprising computers that are physically located. Actually, any computational node can be a component of the distributed system, whether it is a simple electronic device or a person who operates according to his own will. The second property says that each entity in a distributed system can make decisions or execute some actions based on local information, which gives the component the autonomous label. The third property is the main difference between a centralized system and a distributed system. In a centralized system, an oracle, which can be a central controlling unit in the case of computers or a leader in a community, can tell the entities in the system what to do, or how to make a decision. Distributed systems, by contrast, allow more degrees of freedom and the entities are not bound to any oracle. As a connected system, the entities can communicate with others to expand their view of the network, which facilitates better decisions.

1.2 Local Area Networks

A local area network (LAN) is a computer network that interconnects computers in a limited geographical area, such as a school, an office, a factory or a building [4, 10]. A LAN can communicate with other remote LANs (or computers) through dedicated data circuits and such connected LANs form a wide area network (WAN).

In a LAN, each computer can execute computation tasks by itself and no central controlling unit dominates the network. A LAN is normally of the closed type and it can be composed of minimally two computers at home, or thousands of computers in an enterprise. In a LAN, the connected computers can achieve many common functions through the interconnection network, such as sharing files, sending emails, etc.

LANs are ubiquitous in our daily life, due to their scalability, high flexibility, ease of use, and low cost. The main characteristics are:

(1) The LAN covers a small geographical area, or a relatively independent scope;
(2) data can be transferred at a high rate, since LAN can use special transmission media for connection;
(3) the latency is reasonably short, and it has high reliability.

The computers in a LAN are connected directly through cables or by wireless connections. There are many different types of network topology for the interconnection of nodes.

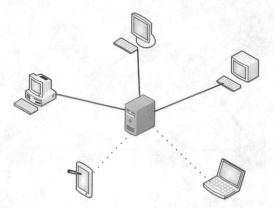

Fig. 1.1 An example of *star* topology in LANs

As shown in Fig. 1.1, in the *star topology*, the computers are connected to a central node (a computer or a server) directly through cables (the solid lines) or via wireless connections (the dotted lines). Any two computers can communicate with each other through the central node and it takes two steps for data transmission. This type of network is easy to manage and the transmission delay is very low. However, the whole network can easily go down if the central node becomes defective or is compromised. Moreover, relying on a central node to relay every single transmission is not preferred in a distributed system.

In Fig. 1.2, in the *ring topology*, all computers are connected in a ring and any two computers can communicate through the interconnected nodes (computers) along the ring. This kind of network is relatively simple, but it is hard to expand and the data transmission rate is relatively low depending on the size of the ring. When any one computer fails, the ring is broken and reconnection is necessary for further communications. The transmission delay is larger than that of the star topology on

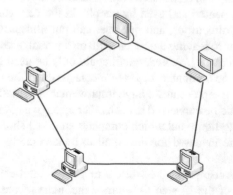

Fig. 1.2 An example of *ring* topology in LANs

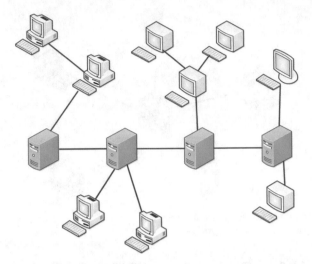

Fig. 1.3 An example of *tree* topology in LANs

average, because two computers may need up to $n/2$ steps to transmit a message (n is the number of computers).

Figure 1.3 shows a tree topology, which is a kind of hierarchical network. There is no loop between any two nodes (computers) and it is easy to expand the network. This kind of networks is generally quite flexible and the failure of the leaf nodes would not affect the whole network.

1.3 Email

Email is another example of distributed systems. Before the ARPANet or the Internet, email had already appeared and used by people. In the very early days, email was implemented as file directories, and one user can put a message in another user's directory. It is similar to leaving a note or a file on a person's desk.

Before the Internet was invented, email could only be used to send messages to other users of the same computer. In 1969, messages could be transmitted between two computers, which represented a huge improvement in email systems. However, when computers could be connected to each other through a common network, there needed to be a way to figure out which computer an email should go to. Therefore, email addresses were invented for the email to be accurately sent to its recipient through the network.

In 1971, Ray Tomlison sent the world's first ever email across a network; and he initiated the use of the @ sign to separate the name of the user and the user's computer. He then succeeded also at sending messages among different users in different computers. This work was quickly adapted to the ARPANet, which started the long and glorious history of emails [34]. By 1974, several hundreds of military users were using email through the ARPANet and it soon became the most popular application among network users.

In 1978, Shiva Ayyadurai took on the challenge of creating a full-scale electronic version of the interoffice paper mail system, and this prototype was officially called "email". In the following year, this email system was adopted by the University of Medicine and Dentistry of New Jersey (UMDNJ), which included Inbox, Outbox, Folder, Attachments, etc. More users, doctors and secretaries were subsequently added to the system. In 1982, Shiva Ayyadurai received the United States copyright for "email".

By 1990s, email systems had grown exponentially in numbers. Microsoft released *Outlook* in 1992; and launched *Hotmail* in 1996, which soon became the most used Internet based system. Following suit, Yahoo, Google and many industrial companies created and released their email systems, which benefited millions of people. Nowadays, email is one of the most important tools in our life, which helps us to connect to even our friends who live far away. This is one of the most successful applications of distributed systems.

1.4 Wireless Sensor Networks

Wireless sensor networks (WSNs) have received a lot of attention from both academic scientists and industrial engineers. As sensors are designed to detect events and changes in the environment, they have been used in many industrial and consumer applications, such as environmental monitoring, health monitoring, battlefield surveillance, intrusion detection, crowd control, etc. [20, 23, 35, 37]. With the ability of transferring data through wireless connections, sensors can send the detected data to a computer or other devices. Wireless sensor networks are composed of such spatially distributed autonomous sensors and these sensors can communicate with nearby nodes (sensors or other devices).

Each sensor in the network may have the following parts:

(1) Radio transceiver, which is used to connect the external antenna;
(2) micro-controller, which controls the sensor itself and decides which tasks to carry out;
(3) battery, which powers the sensor for monitoring, computation and communication.

Some types of sensors may have other specific components, such as a motion part, which enables the sensor to move by itself. With these components, the sensors in a monitored area form a network for data transmission as well as other collaborative functions, which is another example of a distributed system.

Generally speaking, each sensor has a limited communication range since its battery is limited, and two sensors are connected if they are within the communication

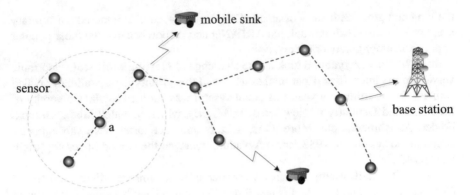

Fig. 1.4 An example of wireless sensor network

range. As illustrated in Fig. 1.4, the dispatched sensors can form a network and they can pass the sensed information to some sink node through multi-hop data transmissions. As shown in the figure, the solid cycle represents a sensor deployed in the environment, and each sensor has a communication range (the dotted cycle centered at sensor a). Any sensor that lies in the range is sensor a's neighbor and they can exchange messages. The network constructed can be used to gather sensed data in the monitored area via information aggregation by the sensors to the base station, or to any nearby mobile sink.

1.5   Cognitive Radio Networks

The wireless spectrum has become a scarce resource due to the rapid advances of wireless technologies and burgeoning growth of wireless handsets. More and more wireless devices, such as laptop computers, mobile phones and various intelligent devices, with increased appetite for wireless bandwidth have been produced.

According to the ABI Research [1], there are more than 10 billion wireless connected devices in 2013 and the number of such devices will exceed 40 billion in 2020. At the same time, new wireless services and demands are making the wireless spectrum even more overcrowded; such services include fast downloading services over wireless frequency bands and high quality video streaming (for webcasting for example) through wireless channels. As a result, wireless spectrum as a resource is in great demand in order to accommodate the billions of wireless devices and increasing number and variety of wireless services.

However, the available wireless spectrum is very limited because the radio spectrum only spans from 3 kHz to 300 GHz [29]. Most of the spectrum has been allocated to license holders or services in advance. The unallocated portion that is free for the billions of wireless devices is becoming very overcrowded, which is known as the spectrum scarcity problem. In this book, we call the spectrum that has already been allocated as the *licensed spectrum* and the unallocated portion the *unlicensed spectrum*.

According to many research results [2, 13–17, 21, 25, 26, 30, 36, 38], the unlicensed spectrum is facing overcrowding due to the increasing numbers of wireless devices and the large number of wireless services, while the utilization of the licensed spectrum continues to remain low. Take for example, the Industrial, Scientific and Medical (ISM) bands which are reserved for industrial, scientific and medical purposes. In recent years, more short-range, low-power communication systems, such as bluetooth devices, mobile phones and wireless computer networks, all use these bands for communication, which causes overcrowding in the ISM bands, and interference problems in these bands become serious and need to be resolved [17]. On the other hand, it is not uncommon that some licensed spectrum is constantly underutilized. For example, the frequencies from 470–698 MHz are allocated to TV broadcasting in the United States, but the utilization of these bands hovers between 15 and 85% [12]. The same phenomenon also occurs in Germany [36], Singapore [21], China [7] and so on.

In order to alleviate the spectrum scarcity problem, a new technique called *Dynamic Spectrum Access (DSA)* was proposed whereby the wireless devices can exploit the licensed spectrum opportunistically by using *cognitive radios* [2, 3]. The concept of cognitive radio was initially proposed in [27, 28] and it has been regarded as an efficient method to improve the utilization of the spectrum [18]. Cognitive radio can be thought of as an intelligent wireless system which can change its parameters based on the interaction with the environment, and it has two distinctive features: cognitive capability and reconfigurability [6, 18, 22, 31]. Cognitive capability means the radio can sense the spectrum to capture the current occupation information about the spectrum, and reconfigurability refers to the ability to sense and access different frequency bands [5]. Through these capabilities, cognitive radios can discover the unused licensed frequency bands that can be used by the wireless devices, which is referred to as *spectrum sensing*.

There are many spectrum sensing techniques, such as energy detection, covariance-based detection, and cooperative spectrum sensing schemes [7–9, 11, 19, 24, 41] and the ultimate objective is to enable the usage of temporally unoccupied spectrum that is referred to as *spectrum hole* or *white space* [18, 40].

As illustrated in Fig. 1.5, the x-axis represents different frequency bands and the y-axis represents the time from 0 to 24h. The solid rectangles with label 1 in the figure represent the frequency bands that are occupied by the licensed users, while the areas with label 2 mean the bands are not occupied all day and the empty rectangles with label 3 stand for the licensed bands that are not in use temporarily. Those areas with label 2 and 3 are called spectrum holes or white space, and wireless devices equiped with intelligent cognitive radios should be able to discover them.

By adopting the DSA technique, cognitive radio networks (CRNs) can be formed to effectively share the licensed spectrum by a large number of unlicensed devices and services. This type of networks can potentially alleviate the spectrum scarcity problem [32, 33, 39].

Fig. 1.5 Spectrum holes:
solid rectangles with label 1
represent the frequency
bands that are occupied;
areas with label 2 mean the
unused bands; *empty
rectangles* are the frequency
bands that are not used
temporally

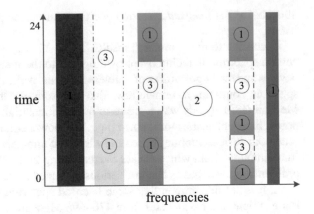

Similar to the division betweeen licensed spectrum and unlicensed spectrum, there
are two types of users coexisting in the network, primary users (PUs), i.e. licensed
users who own the licensed spectrum, and secondary users (SUs), i.e. unlicensed users
that are equipped with cognitive radios to sense and access the licensed spectrum
opportunistically.

As depicted in Fig. 1.6, both PUs and SUs are present in the network, and the SUs
can detect the licensed spectrum, and utilize the unoccupied licensed spectrum to
communicate with each other. The figure depicts a network consisting of five SUs

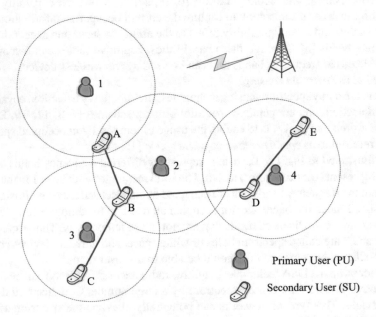

Fig. 1.6 An example of cognitive radio network

and four PUs that are distributed in a small geographical area. As illustrated in the figure, the dotted cycle of each PU represents the PU's interference range and any SU within the range thus cannot access the channels occupied by the PU.

The SUs have the ability to discover the unused licensed spectrum and to communicate with other wireless devices through it. As soon as the PUs need to use the spectrum again, the SUs have to stop the communication process and their cognitive radios may seek some other unoccupied frequency bands to continue the communication. This new paradigm can improve a spectrum's utilization by sharing it with unlicensed devices while guaranteeing high priority access to the frequency bands by the licensed users. Therefore, real efforts have been made to promote the development of CRN. For example, IEEE 802.22 [8] and IEEE 802.11 *af* [1] are two ongoing standards in spectrum sharing; the United States is extending the TV white space into the 3550–3700 MHz US Navy radar band, while Europe has been pursuing an authorized shared access licensing model [29].

1.6 Telephone Networks

A telephone network is a telecommunication network which benefits two or more parties making a telephone call. There are two common types of telephone networks:

(1) The public switched telephone network (PSTN), which is a landline network and the telephones are connected by wires;
(2) wireless networks where the telephones are mobile and they can communicate through wireless media.

The public switched telephone network consists of telephone lines that connect telephones, fiber optic cables for data transmission, microwave transmission links,

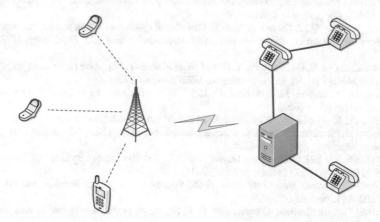

Fig. 1.7 An example of telephone network

undersea cables for long-haul connection, and switching centers. These components enable the telephones to communicate with each other.

The wireless network is usually composed of mobile telephones and they may transmit communication data through the wireless spectrum. Nowadays, the number of mobile phones is increasing so rapidly that the wireless spectrum space for their communication is becoming extremely crowded. These mobile phones equipped with intelligence can make decisions locally and they constitute a distributed system.

As illustrated in Fig. 1.7, both mobile phones and landline telephones are fully connected in the network, and any two telephones can reach each other through the network. When we dial a number, the network will find the target terminal phone and pick a communication path through the inter-connected components. This process is similar to emails, which like emails is also an important and frequently used application in our daily lives.

References

1. ABI Research. https://www.abiresearch.com.
2. Akyildiz, I. F., Altunbasak, Y., Fekri, F., & Sivakumar, R. (2004). Adaptnet: adaptive protocol suite for next generation wireless internet. *IEEE Communications Magazine, 42*(3), 128–138.
3. Akyildiz, I. F., Lee, W., Vuran, M., & Mohanty, S. (2006). Next generation//dynamic spectrum access//cognitive radio wireless networks: a survey. *Computer Networks, 50*(13), 2127–2159.
4. Andrews, G. R.(2000) *foundations of multithreaded, parallel, and distributed pragramming*. Addison-Wesley.
5. Bahl, P., Chandra, R., Moscibroda, T., Murty, R., & Welsh, M. (2009). White space networking with wi-fi like connectivity. In *SIGCOMM*.
6. Chen, K.-C., Peng, Y.-J., Prasad, N., Liang, Y.-C. & Sun, S. (2008). Cognitive radio network architecture: part I - general structure. In *International Conference on Ubiquitous Information Management and Communication (ICUIMC)*.
7. Chen, D., Yin, S., Zhang, Q., Liu, M., & Li, S. (2009). Mining spectrum usage data: A large-scale spectrum measurement study. In *Proceedings of the 15th Annual International Conference on Mobile Computing and Networking*.
8. Chen, Y., Zhao, Q., & Swami, A. (2009). Distributed spectrum sensing and access in cognitive radio networks with energy constraint. *IEEE Transactions on Signal Processing, 57*(2), 783–797.
9. Chen, Z. & Qiu, R. C.(2010). Prediction of channel state for cognitive radio using higher-order hidden Markov model. In *Proceedings of IEEE Southeast Con*.
10. Clark, D. D., Pogran, K. T. & Reed, D. P.(1978). An introduction to local area networks. In *Proceedings of the IEEE*.
11. Datla, D., Rajbanshi, R., Wyglinski, A. M., & Minden, G. J. (2009). An adaptive spectrum sensing architecture for dynamic spectrum access networks. *IEEE Transactions on Wireless Communications, 8*(8), 4211–4219.
12. ETSI. EN 301 598 White Space Devices (WSD); Wireless Access Systems Operating in the 470 MHz to 790 MHz Frequency Band, 2012.
13. Federal Communications Commission. FCC Spectrum Policy Task Force Report ET Docket No. 02–155, November, 2002.
14. Federal Communications Commission. FCC Notice of proposal rulemaking and order, ET Docket No. 03-322, 2003.
15. Federal Communications Commission. (FCC) manages and regulates all domestic non-federal spectrum use (47 USC 301), 2004.

16. Flores, A. B., Guerra, R. E. & Kightly, E. W.(2013) IEEE 802.11af: A standard for tv white space spectrum sharing. In *IEEE Communications Magazine*.
17. Fragkiadakis, A. G., Tragos, E. Z., & Askoxylakis, I. G. (2013). A survey on security threats and detection techniques in cognitive radio networks. *Communications Survey Tutorials*, *15*(1), 428–455.
18. Haykin, S. (2005). Cognitive radio: Brain-empowered wireless communications. *IEEE Journal on Selected Areas in Communications*, *23*(2), 201–220.
19. Haykin, S., Thomson, D. J., & Reed, J. H. (2009). Spectrum sensing for cognitive radio. *Proceedings of the IEEE*, *97*(5), 849–877.
20. Hart, J. K., & Martinez, K. (2006). Environmental sensor networks: a revolution in the earth system science? *Earth-Science Reviews*, *78*, 177–191.
21. Islam, M. H., Koh, C. L., Oh, S. W., Qing, X., Lai, Y. Y., Wang, C., Liang, Y.-C., Toh, B. E., Chin, F., Tan, G. L., & Toh, W.(2008) Spectrum survey in Singapore: 180 Occupancy measurements and analyses. In *Proceedings of 3rd International Conference on Cognitive Radio Oriented Wireless Networks and Communications*.
22. Jondral, F. K.(2005) Software-defined radio-basic and evolution to cognitive radio. *EURASIP Journal on Wireless Communication and Networking*.
23. Ma, Y., Richards, M., Ghanem, M., Guo, Y., & Hassard, J. (2008). Air pollution monitoring and mining based on sensor grid in London. *Sensors*, *8*(6), 3601–3623.
24. Ma, J., Li, G. Y., & Juang, B. H. (2009). Signal processing in cognitive radio. *Proceedings of the IEEE*, *97*(5), 805–823.
25. McHenry, M. A. (2005). *NSF spectrum occupancy measurements project summary*. Technical Report: Shared Spectrum Company.
26. McHenry, M. A., Tenhula, P. A., McCloskey, D., Roberson, D. A., & Hood, C. S.(2006). Chicago spectrum occupancy measurements & analysis and a longterm studies proposal. In *Proceedings of the First International Workshop on Technology and Policy for Accessing Spectrum*.
27. Mitola, J, I. III & Maguire, G. Q. (1999). Cognitive radio: making software radios more personal. *IEEE Personal Communications*, *6*(4), 13–18.
28. J. Mitola III (2000). Cognitive radio: an integrated agent architecture for software defined radio. Ph.D. Thesis, KTH Royal institute of technology.
29. Spectrum Management. http://en.wikipedia.org/wiki/Spectrum_management.
30. Stevenson, C. R., Chouinard, G., Lei, Z., Hu, W., Shellhammer, S. J., & Caldwell, W. (2009). IEEE 802.22: The first cognitive radio wireless regional area network standard. *IEEE Communications Magazine*, *47*(1), 130–138.
31. Thomas, R.W., DaSilva, L.A., MacKenzie, A.B.,(2005). Cognitive networks. In *IEEE DySPAN*.
32. Thomas, R. W., Friend, D. H., DaSilva, L. A., & MacKenzie, A. B. (2007). Cognitive networks: adaptation and learning to achieve end-to-end performance objectives. *IEEE Communications Magazine*, *44*(12), 51–57.
33. Thomas, R. W.,(2007) Cognitive networks. Ph.D. dissertation, Virginia Polytechnic Institute and State University.
34. Tomlinson, R.(2014). The first network Email. http://openmap.bbn.com.
35. Vasilescu, I., Kotay, K., Rus, D., Dunbabin, M., & Corke, P.(2005). Data collection, storage, and retrieval with an underwater sensor network. In *Proceedings of the 3rd international conference on Embedded networked sensor systems (SenSys)* New York, NY, USA.
36. Wellens, M., Riihijarvi, J., & Mahonen, P. (2008) Evaluation of spectrum occupancy using approximate and multiscale entropy metrics. In *Proceedings of 5th IEEE annual communications society conference on sensor, mesh and ad hoc communications and networks workshops*.
37. Willkomm, D., Machiraju, S., Bolot, J. Wolisz, A.(2008). Primary users in cellular networks: A large-scale measurement study. In *Proceedings of 3rd IEEE Symposium on New Frontiers in Dynamic Spectrum Access Networks*.
38. Werner-Allen, G., Lorincz, K., Welsh, M., Marcillo, O., Johnson, J., Ruiz, M., et al. (2006). Deploying a wireless sensor network on an active volcano. *IEEE Internet Computing*, *10*(2), 18–25.

39. Wyglinski, A. M., Nekovee, M., Hou, T. (2010). *Cognitive radio communications and networks: principles and practice*. Academic Press, 2010.
40. Yuan, Y., Bahl, P., Chandra, R., Chou, P. A., Ferrel, J. I., Moscibroda, T., Narlanka, S. & Wu, Y. (2007). KNOWS: Kognitiv networing over white spaces. In *DySpan*.
41. Zhao, Q., & Sadler, B. M. (2007) A Survey of Dynamic Spectrum Acess (signalprocessing, networking and regulatory policy). *IEEE Signal Processing Magazine*.

Chapter 2
Distributed Computing

Abstract Distributed computing studies the theory and methods to solve computational problems in distributed systems. There are many interesting and important problems that can be solved efficiently in distributed systems, such as data gathering in wireless sensor networks, computing graph properties, and leader election in a distributed system. In this chapter, we introduce the elementary concepts about distributed computing and some of the important components of distributed computing. In Sect. 2.1, we introduce the concept of distributed computing and present an example to illustrate it. Then, we present the communication models that are commonly utilized in Sect. 2.2, show the incompleteness of information in Sect. 2.3, and discuss the aspect of timing which plays an important role in distributed computing in Sect. 2.4.

2.1 What is Distributed Computing?

Distributed computing studies how to solve computational problems in distributed systems or environments. Generally speaking, an entity in a distributed system can compute its tasks completely locally against its "individual" goals. An analogy would be that each individual has his own will and sentiments about certain public events. However, these entities can also cooperate to solve global computational problems which are tough or impossible for a single entity to handle, even though they may not know the others' information. For example, all individuals can contribute their strength in crowdsourcing to provide some common needed services or to achieve some common goal [1, 2, 6].

We use a simple example of wireless sensor networks to explain distributed computing. As illustrated in Fig. 2.1, 9 sensors are deployed in the environment to detect the temperature. The goal is to find out the highest temperature in the area at the base station through the sensors. As shown in the figure, the sensors form a wireless network where any two sensors are supposed to be connected if they are within each other's range of communication. Each sensor node can detect the temperature locally and then share this local data by communicating with nearby neighboring sensors. The 9 sensors can sense different temperature data and after receiving the others' data

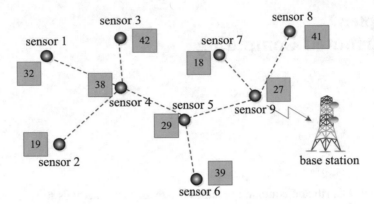

Fig. 2.1 An example of computing the maximum temperature in a wireless sensor network

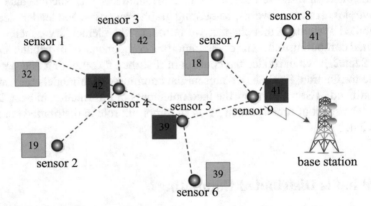

Fig. 2.2 An example of update dating when we are to compute the maximum temperature through the wireless sensor network

(the temperature data), each sensor can compute which temperature is the highest among all neighbors and then update the data.

In Fig. 2.2, sensor 4 will update the local highest temperature data to 42 since it can collect four values {19, 31, 38, 42}. Similarly, sensor 5 will update the local temperature to 39 and sensor 9 to 41. However, having updated its data, a sensor would communicate its updated data with the neighbors again. For example, when sensor 4 has updated its max value to 42, it will also send the data to sensor 5, which would also update the local highest temperature data to 42. After sensor 5 has updated to 42, it will then send the information to sensor 9 which will finally send the data to the base station. In this fashion, all sensors can cooperate to compute the highest temperature and inform the base station.

Distributed computing is quite different from centralized computing where all entities know the global information ahead of time. Considering the above example, if all nodes know the others' temperature data in advance, they (including sensor 9) can easily find out the global highest temperature. In distributed computing, although each entity focuses on its private computation tasks, through reasonable communication (data transmission), the private computations of the entities can be combined to realize a global computational task, which should produce the same result as centralized computing, but with the extra cost of longer elapsed time.

Distributed computing has the following three traits. The first one is the communication model, which dictates how data are to be transferred between two entities; the second one is the incompleteness of information, which reveals to what extent an entity can know about the whole system; the third one is synchrony and timing, which indicates when the entity starts its computation and how the computation should be paced.

2.2 Communication Model

In a distributed system, two connected entities can exchange and share their information. There are two commonly used communication models:

(1) *Message passing model (MPM)*: an entity can send data to its neighbors through the connecting edges;
(2) *shared memory model (SMM)*: the entities can use some kind of common memory to perform data transmission.

In the message passing model, the communications between two entities are explicit. When one entity wants to share information, it can transmit the data through the edges connecting the neighboring entities. A large number of message passing models have been proposed in the past; here, we describe two of them.

One type of message passing model is called *point-to-point* communication, which allows direct information transmission between a specific pair of entities. For example, in Fig. 2.3, the edges with both start and end arrows represent bidirectional communication connections (or channels) between two entities, where bidirectional communication means one node (entity) can both send and receive information to/from the other node of the edge. In the figure, nodes a and b can both share information with the other (the red rectangle represents the message). However, some systems may use just unidirectional communication connections, i.e. one node can only send or receive from the other node. We use an edge with only one arrow in the figure to represent a unidirectional connection. As shown in the figure, node c has a directed edge to node a, which means node c can send its message to node a, but node a cannot share its information with node c. Edges from node d to c, node e to node d, node e to node g, and node h to node f are likewise unidirectional edges.

From the figure, node a can get node b's and node c's data through the edges, but it cannot receive both pieces of data at the same time. Point-to-point communication means that the data transmission process happens between a pair of neighbors, but one node cannot receive data from multiple senders simultaneously. Through continuous communication, node b can also receive node c's information through node a, but

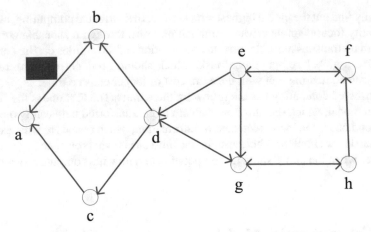

Fig. 2.3 An example of message passing model

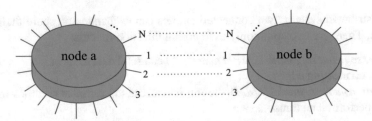

Fig. 2.4 An illustration of message passing model with external ports for connection

it takes longer time and we say the information is delayed. However, node c cannot get node b's information through the depicted topology.

Another type of message passing model is called *broadcast*, where one node can send its information to *all* the neighbors simultaneously. Its main difference from the point-to-point communication model, where a message can be only sent to one receiver at a time, is that it allows multiple recipients at the same time. For example, in Fig. 2.3, node d can broadcast its message and nodes b, c, g can receive this message simultaneously. This broadcast model has many important applications, such as flooding in network construction [4, 5], and fast message propagation [7].

The first type of communication model is widely adopted since it reveals the connection patterns directly. However, there is a more refined model of communication, as shown in Fig. 2.4. The edges in the graph (Fig. 2.4) represent direct connections between two nodes, which also means the two nodes are relatively close to each other; but they still need to rely on a connection channel for communication. Suppose each node has a number of *ports*, i.e. external connection points, and every communication channel connects two ports of two connected nodes. When node a tries to send a message to node b, it should load the data onto an appropriate port, such as port 1; when the message arrives at node b, it will be stored in b's local buffer. However, if node b does not choose port 1 or the connection channel between the two ports cannot be established, node b cannot receive the message. This book focuses on the rendezvous problem in distributed systems, which is the process of establishing a common communication link (connection) between two connected external ports of the communicating entities.

2.3 Information Incompleteness

In executing computational tasks in a networked system, centralization helps solve the problems from a global view where each node in the system has full knowledge of all relevant information about the tasks. Therefore, each node will achieve the same result with the full knowledge (we do not consider randomized algorithms here, which may lead to different results even with same input). The completeness of information is equivalent to a full input to any problem in such a centralized setting.

However, centralized setting is hard to implement is real distributed systems. For example, there are more and more mobile phones becoming active nowadays, and it is costly to construct a static network and inform all users about the new phones' information and the constructed topology. In distributed systems, every entity needs to cope with the fact that only partial information of the system is available. This is equivalent to the situation where only partial input to a problem is available, and the user has to execute its task with its stored or information obtainable from the surrounding. Moreover, each entity may not even be aware of who the other participating entities are, where the computation begins, and which stage of the computation the others are currently at. These uncertainties lead to difficulties in coordinating the joint computation of the entities of a common task.

In some practical applications or computational tasks, the entity of a distributed system may not need to have the full knowledge about the system. For example, if each entity should compute the number of connected neighbors dynamically (since some entities may join in or leave the system at any time), it only needs to find out the active entities within its communication range. Actually, full system information does not help perform such a task. Therefore, the entity may not need to know all the outsiders' information, and collectively the entities can also work well with only local information in solving many computational problems.

There are a variety of models that govern concern the topological knowledge. One typical model is known as anonymous system, where all entities (or nodes) are indistinguishable and they have no identification labels. Moreover, each node knows nothing about the topology of the network. This model is at the extreme end of the spectrum, which makes distributed computing difficult. For example, it is hard for a node to find out whether it has sent a message to all neighbors in a point-to-point communication model, since it cannot tell whether two "different" end-points are the same node. A more realistic model assumes that each node is assigned an unique identifer and the node knows the identities of its neighbors. For example, a computer or a mobile phone has a unique MAC address, which can be discovered by others.

Some models assume a node know even more about the system. For example, each node may know the k-hop network topology, which means it is able to find out the nodes that are reachable within k hops. This subnetwork information can help solve some problems to a certain extent. When k becomes larger, more information can be obtained by each node. The most powerful model assumes each node can have the complete topological knowledge of the system, which degenerates to the centralized setting; centralized setting is impractical and hard to implement.

2.4 Timing and Synchrony

In distributed computing, timing is a subtle concept that deals with when the entities may execute their tasks or computational steps [3]. In our normal living, we have a global clock that tells the time and all the entire world agrees to the same rules for defining time. However, in a distributed system, timing is hard to coordinate and a global clock is hard to implement if it is to to be used by all distributed entities.

We mention two models that have to do with timing: synchronous model and asynchronous model.

In the *synchronous model*, all entities in the distributed system share a global clock, which indicates the exact times of the events in the system. Time can be considered as divided into slots of equal length (the analogy is that we use "second" as the elementary unit in physical clocks), and each entity should execute the following three steps in each slot:

(1) Receive messages from (some of) the neighbors;
(2) execute local computation based on its local status and the received messages;
(3) send messages to (some of) the neighbors.

The time cost of local computation on each entity is assumed to negligible compared to the message transmissions. Therefore, in the model, an entity only needs to wait for its message and then send out a computed message. This model satisfies an important property: if entity a sends a message to entity b in slot t, the message must be received by entity b before or in slot $t + 1$. Thus, all entities' activities can be regarded as driven by a global clock.

However, in the *asynchronous model*, messages are not guaranteed to be transmitted to the other entity timely. All entities do not access the global clock and they have to decide on their own actions. Generally speaking, messages sent from one entity to another will arrive within some finite but unpredictable time. Therefore, one cannot rely on the elapsed time to deduce whether a message was sent from a neighbor or not. Thus, the algorithms for this model are always event driven, i.e. the entity will execute its local computations when a message is received, or when some local memory has changed. Therefore, the execution steps are as follows:

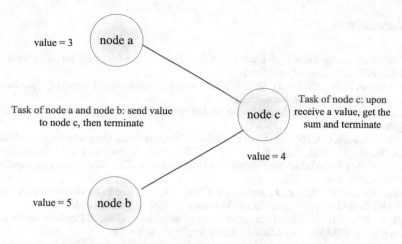

Fig. 2.5 An example of distributed computing upon received messages

(1) Wait for an event, where the event could be receiving messages from (some of) the neighbors, or the local memory has changed;
(2) execute local computations based on its local memory and the received messages;
(3) trigger an event, such as sending messages to (some of) the neighbors, or change the local memory.

Clearly, the entities' computations could be affected by the messages' arrival times. However, it is impossible to rely on the ordering of the arrived messages for executing local computations.

For example, in Fig. 2.5, node a and node b are connected to node c. The tasks for node a and node b are to send its local value to node c, and then terminate. Meanwhile, the task of node c is to add the received values and update its local value; and then, it will terminate. In the asynchronous model, the messages could be delayed by different reasons. Suppose node a sends the value to node c earlier than node b, if the messages can arrive at node c timely and sequentially, node c will get node a's value and update its local value to $3 + 4 = 7$. However, in the asynchronous system, node b's message may arrive earlier at node c and it will update the value to $5 + 4 = 9$, which leads to different results.

Therefore, timing plays an important role in distributed computing. Some asynchronous models assume the entities start the algorithm in different time slots, but the messages are guaranteed to be received in the next time slot. Therefore, in the above figure, if one node sends its local value earlier than the other, for example, node a sends the data $\Delta > 0$ time slots earlier than node b, node c will update the value to $3 + 4 = 7$ timely, and then terminate. In this book, we design efficient distributed algorithms for both synchronous and asynchronous scenarios; we use exactly this type of asynchronous model which we will introduce later.

References

1. Doan, A., Ramakrishnan, R., & Halevy, A. Y. (2011). Crowdsourcing systems on the world-wide web. *Communications of the ACM* 54(4).
2. Huberman, B. A., Romero, D. M. & Wu, F. (2009). Crowdsoucring, attention and productivity. *Journal of Information Science*, 35(6).
3. Lamport, L.(1978). Time, clocks, and the ordering of events in a distributed system. *Operating System*.
4. Lim, H. & Kim, C. (2001). Flooding in wireless ad hoc networks. *Computer Networks*, 24(3–4).
5. Liu, H., Jia, X., Wan, P.-J., Liu, X., & Yao, F. F. (2001). A distributed and efficient flooding scheme using 1-hop information in mobile ad hoc networks. *IEEE Transactions on Parallel and Distributed Systems*, 18(5).
6. Morschheuser, B., Hamari, J., Koivisto, J. (2016). Gamification in crowdsourcing: A Review. In *49th Annual Hawaii International Conference on System Sciences (HICSS)*
7. Ye, S., Wu, S. F. (2010). Measuring message propagation and social influence on twitter. In *International Conference on Social Informatics*, Springer, Berlin

Chapter 3
Rendezvous Theory

Abstract In distributed computing, the entities of the system are assumed to be able to communicate with each other and predominantly they adopt the message passing model (point-to-point model, broadcast model or others) for communication. However, many works abstracted away the detailed communication scheme and assume that two entities can transmit messages to each other if they are connected in the network. Actually, each entity has many external ports for communication as described in Chap. 2, and two neighbors can only communicate with each other if indeed a communication link can be constructed between the ports they choose. Rendezvous is referred to as the process of two entities choosing the connected ports for link establishment and data transmission. In this chapter, we introduce the rendezvous problem, how it is applied in distributed computing, and in which areas. In Sect. 3.1, we give the definition of the rendezvous problem, present rendezvous examples in distributed systems including multichannel wireless networks and cognitive radio networks in Sects. 3.2 and 3.3. Afterwards, we define rendezvous in distributed system in Sect. 3.4 and present some simple distributed algorithms in Sect. 3.5.

3.1 What is the Rendezvous Problem?

To begin with, consider the following scenario: two young people want to meet each other at some place (such as a supermarket or a large park), but they cannot find each other since it is a huge area and they do not have the specific location to meet. Rendezvous, which means getting together at exactly the same place, asks how they can meet quickly. Obviously, each person can choose between waiting in a fixed place or looking for the other person by searching the different places; obviously, if they both wait, they will never rendezvous. The objective is to design efficient strategies for them to achieve rendezvous as quickly as possible under the assumption that neither one knows the other person's strategy and they do not make any plans beforehand.

This rendezvous problem was first officially introduced in [1] and the continuous version of the problem was formalized in [2]. Following the pioneering work, the rendezvous problem has been studied and applied in many research areas such as

Fig. 3.1 An example of telephone coordination problem

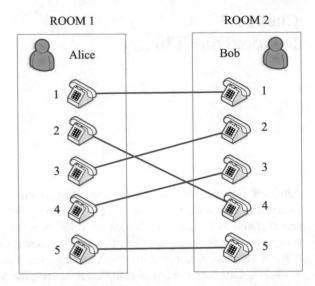

graph theory [3, 4, 6, 18, 33], control theory [16, 39], and mobile robots [19, 28, 38]. In recent years, the rendezvous problem has been applied to wireless networks in designing efficient multichannel MAC protocols, where the wireless devices can access multiple orthogonal channels for communication in order to increase capacity and throughput [34, 35]. In addition, the rendezvous problem for Cognitive Radio Networks (CRNs) has been drawing a lot of attention in the last decade, since CRNs are the most typical multichannel wireless networks and they are the most promising prototype for the next generation wireless networks.

We first introduce a simple example of the rendezvous problem, which is known as the *telephone coordination problem* [1]. As depicted in Fig. 3.1, two people (Alice and Bob) are isolated in two rooms and they can only communicate through the telephones placed in each room. However, they do not know how these telephones are connected and neither person knows when the other will pick which telephone. If both users pick telephone 1 at the same time by luck, they can hear from each other and we call that a *rendezvous*. However, if Alice happens to pick telephone 2 while Bob picks telephone $k \neq 4$, they will not hear anything from the other room and they have to choose between holding the telephone or picking another one, which is similar to the starting scenario in this section.

3.2 Rendezvous in Multichannel Wireless Networks

In multichannel wireless networks (MWN), as in Fig. 3.2, the wireless device is like the person in the example of telephone coordination problem (Fig. 3.1), and the telephones placed in the rooms correspond to the wireless channels that the devices

Fig. 3.2 Rendezvous for
multichannel wireless
networks

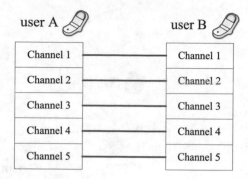

can use for communication, and the connection patterns between the telephones
can be thought of as the real physical connections among the channels. Similar to
telephone coordination, if both devices choose the same channel at the same time,
rendezvous is achieved and they can communicate with each other. The difference
between telephone coordination and multichannel wireless networks is that the wireless devices have the same labels as the wireless channels and rendezvous protocols
can be designed based on the shared labels.

3.3 Rendezvous in Cognitive Radio Networks

Cognitive Radio Network (CRN) has been hailed as a promising network scheme
for solving the spectrum scarcity problem, and rendezvous for CRNs is defined as
"the process of one smart or cognitive radio finding another in the spectrum band
of interest" [41]. some works also refer to the problem as *neighbor discovery* [7,
9, 52]. In [7, 31], rendezvous is regarded as "the process of two or more radios of
users to meet and establish a link on a common channel", which is more formal and
intuitive. Actually, establishing a communication link on a channel involves many
detailed implementation steps, such as beaconing and handshaking [41]; we refer
to rendezvous process as the process of *choosing the same common channel at the
same time*. Although CRN is one special type of multichannel wireless networks, the
rendezvous problem in CRN is quite different from both the telephone coordination
problem and rendezvous in multichannel wireless networks.

Take Fig. 3.3 as an example, and assume there are six licensed channels {1, 2, 3, 4,
5, 6} for both primary users (PUs), i.e. the users who own the licensed spectrum, and
secondary users (SUs), i.e. the users who can only use the unlicensed spectrum; PU
1 occupies channels {3, 4}, PU 2 occupies channels {6}, PU 3 occupies channels
{1, 2}, and PU 4 occupies channels {2, 3}, and then the network can be thought of
as in Fig. 3.3. Due to the PUs' occupancy of the licensed channels, not all channels
are available for the SUs and this is a huge difference from both the telephone
coordination problem and multichannel wireless networks. As shown in the figure,

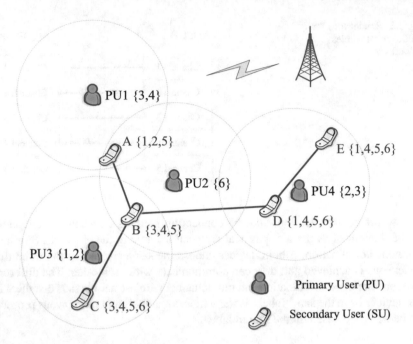

Fig. 3.3 Rendezvous example for cognitive radio network

users *A* and *B* can rendezvous on channel 5, users *B* and *C* can rendezvous on channels {3, 4, 5}, users *B* and *D* can rendezvous on channels {4, 5}, and users *D* and *E* can rendezvous on channels {1, 4, 5, 6}. The objective of rendezvous is to design channel accessing algorithms such that every pair of neighboring users (SUs) can access the same available licensed channel at the same time as quickly as possible. Compared with the traditional rendezvous problem, rendezvous in CRN is more challenging for the following reasons:

(1) The wireless devices (SUs) can only access the unused licensed frequency bands (channels) after the spectrum sensing stage, which means not all channels are available to choose from;
(2) traditional multichannel wireless networks assume each frequency band has a fixed label that all SUs are aware of. However, different entities or service providers may have different labels even for the same frequency band in CRN and is no standard for everyone to subscribe to;
(3) different wireless devices may have different capabilities to sense the licensed spectrum, which means the SUs can only sense a portion of the spectrum and they may have only a small common fraction of the total spectrum available for communication;
(4) the status of the licensed spectrum may vary according to the licensed users' (PUs') activities, which implies the sensed available channels can change dynamically.

The first challenge comes from the fact that different SUs may have different available channels to use, which makes the rendezvous problem harder. For example, if both SUs choose channel 1 in Fig. 3.2, they cannot rendezvous if the channel is occupied by nearby PUs, which implies they do not have the chance to use the channel.

The second challenge is due to frequency bands being labeled differently or locally by different SUs. For example, the SUs in Fig. 3.2 cannot rendezvous on channel 1 if they represent different frequency bands. Many extant works assume the labels are fixed and all devices share the same information. The study of a practical problem called *oblivious blind rendezvous* was proposed in [5, 23] where the SUs have no common labels of the channels.

The third challenge arises from the different sensing capabilities of different SUs, and they constitute the so called *Heterogeneous Cognitive Radio Network (HCRN)* [37, 49, 50].

The forth challenge reflects the impact of PUs on the SUs; whenever the PUs occupy the spectrum, the SUs have to release it and exploit new opportunities for rendezvous. Therefore, the status of the licensed channels varies temporally and rendezvous needs to be achieved in a dynamic way.

3.4 Rendezvous in Distributed Systems

From the above examples of the rendezvous problem, we now have the essential meaning of rendezvous in distributed systems: to construct a communication link from the entities' connected ports (or chosen channels) for message transmission.

Considering two neighboring entities in a distributed system, suppose each one has N external ports (which corresponds to N telephones in the telephone coordination problem). Suppose these external ports are labeled with distinguishable identifiers, and some external ports between two neighbors are connected. For example, Fig. 2.4 shows the connections between the neighbors. For node a and b, they can only transmit data to each other when they choose the connected ports, even if they are connected in the network.

Rendezvous in distributed systems is similar to the problem in cognitive radio networks, where each entity (node) does not know the other's information, such as how many ports are connected, how the ports are labeled, etc. All entities have to find out their connections with neighbors on the basis of its local information, i.e. the external ports and perhaps some distinguishable identifiers (such as ID, MAC address, etc.).

Clearly, rendezvous plays an important and fundamental role in distributed systems, which reveals the underlying communication principles. This process can be adopted to construct wireless networks, discover neighbors for autonomous robots, find the existence of other sensor nodes in sensor networks, etc. In this book, we focus on introducing some basic principles and elementary methods of designing efficient rendezvous algorithms, and presenting some elegant results for the field.

3.5   Distributed Rendezvous Algorithms

As very few works have studied the rendezvous process of choosing the connected ports in distributed systems, we mainly highlight some results of rendezvous in some specific systems. Combining these results, we then introduce how to design efficient rendezvous algorithms in distributed systems.

3.5.1   Distributed Telephone Coordination Algorithms

The *telephone coordination problem* seeks to minimize the elapsed time until two isolated individuals can pick the pair of connected telephones. We formally introduce the problem as follows:

Two persons (Alice and Bob) are isolated in two rooms, and there are N telephones in each room. These N telephones are pairwise connected but no one knows how they are connected. Alice and Bob can only communicate with each other by picking up the connected telephones at the same time. Suppose they try round after round, where each person can choose one telephone in each round. If their chosen telephones are connected, rendezvous is achieved and they can communicate. Each person does not know the other person's information (such as identifiers, or telephone chosen strategy); the problem seeks to minimize the expected time (i.e. number of rounds) to achieve rendezvous.

For simplicity, assume the telephones of each person is labeled as $\{1, 2, \ldots, N\}$ randomly, and telephone i of Alice is connected to certain telephone j of Bob (which is not known to the two users). A naïve but reasonable strategy is to pick the telephone randomly in each round. We show that such an algorithm uses expected N rounds for two people to rendezvous.

Lemma 3.1 *Random picking strategy has expected time to rendezvous $ETTR = N$ for two people.*

Proof Let r_t be the event that Alice and Bob pick a pair of connected telephones in the t-th round. Since each person chooses the telephone randomly,

$$Pr(r_t) = \frac{1}{N} \tag{3.1}$$

which computes the probability of event r_t happening. Let r_t' be the event that the users can rendezvous in the t-th round for the first time, then

$$Pr(r_t') = Pr(\overline{r_1} \bigcap \overline{r_2} \bigcap \ldots \bigcap \overline{r_{t-1}} \bigcap r_t) = \left(1 - \frac{1}{N}\right)^{(t-1)} \cdot \frac{1}{n} \tag{3.2}$$

Thus, we derive the expected time to rendezvous as:

$$ETTR = \sum_{t=1}^{\infty} t \cdot Pr(r_t')$$

$$= \sum_{t=1}^{\infty} t \cdot \left(1 - \frac{1}{N}\right)^{(t-1)} \cdot \frac{1}{N}$$

$$= N$$

Following this expected rendezvous time, we can conclude that the users can achieve rendezvous within $O(N \log N)$ rounds with high probability.

Lemma 3.2 *Random picking strategy guarantees rendezvous in $O(N \log N)$ rounds for two people with high probability.*

Proof As shown in Lemma 15.2, the probability to rendezvous in each round t is $Pr(r_t) = \frac{1}{N}$. Since both persons select their telephones randomly in each round, r_t, r_t' are independent for any $t \neq t'$. Therefore, the probability that they do not rendezvous in $cN \log N$ (c is a constant) rounds is bounded by:

$$Pr(\overline{r_1} \bigcap \overline{r_2} \bigcap \cdots \bigcap \overline{r_{cN \log N}}) = \left(1 - \frac{1}{N}\right)^{cN \log N} \tag{3.3}$$

When $N \to \infty$,

$$Pr(\overline{r_1} \bigcap \overline{r_2} \bigcap \cdots \bigcap \overline{r_{cN \log N}}) = e^{-c \log N} = \frac{1}{N^c} \tag{3.4}$$

and thus rendezvous happens in $O(N \log N)$ time slots with high probability $1 - \frac{1}{N^c}$.

The random picking strategy is simple. A better algorithm is proposed in [6], which is called the Anderson-Weber (AW) strategy. This algorithm works as follows:

(1) Choose a random value $i \in [1, N]$ and pick the telephone with label i in the first round;
(2) choose a constant value $p \in [0, 1]$ and pick the telephone with label i for the next $N - 1$ rounds with probability p, or pick the telephones in the next $N - 1$ rounds according to a random permutation of label set $\{1, 2, \ldots, i - 1, i + 1, \ldots, N\}$ (with probability $1 - p$);
(3) If rendezvous does not happen, repeat the second step.

In [6], the AW strategy is proved to be better than random picking strategy and the expected time to rendezvous ($ETTR$) is about $0.829N$. We will introduce the details in Chap. 15.

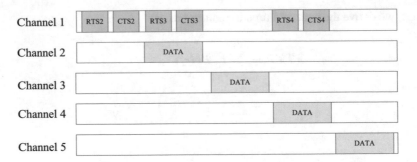

Fig. 3.4 Dedicated Common Control Channel based rendezvous protocol

3.5.2   Distributed Rendezvous Algorithms for Multichannel Networks

In traditional wireless networks, there are mainly two types of multichannel MAC protocols for rendezvous: dedicated common control channel (CCC) based rendezvous [25, 47, 48] and channel hopping based rendezvous [8, 13, 42, 43, 45].

The intuitive idea of *dedicated CCC based rendezvous protocols* is to schedule the requests of establishing a link for communication in a dedicated fixed channel. As depicted in Fig. 3.4, channel 1 is the control channel via which the users make an agreement for rendezvous.

For example, one device (we denote it as user *A* for simplicity) wants to transmit data to user *B* on channel 2; it sends a request-to-send (RTS) packet on channel 1 and user *B* responses with a clear-to-send (CTS) packet if it is ready for communication. Then the users utilize channel 2 for communication and no other users can occupy the channel until they finish the transmission. Dynamic Channel Allocation (DCA) [47], Dynamic Channel Allocation with Power Control (DCA-PC) [48], and Dynamic Private Channel (DPC) [25] are some typical examples of this method.

The main advantage of the approach is that it can ensure all users be aware of the status of each channel through communication on the dedicated control channel. Moreover, time synchronization is not needed, which means the users can always achieve rendezvous via the control channel no matter when they want to communicate with others. However, there are several disadvantages:

(1) If the number of channels that can be used is small, the efficiency of utilizing the channels for communications decreases since the dedicated control channel cannot be used for transmission between two specific users;
(2) the dedicated control channel can be easily congested if the number of users is large;
(3) once an attacker finds out the frequency of the dedicated control channel, it can attack and block the channel, which results in no more rendezvous subsequently.

In order to overcome the first disadvantage, a new method called *split phase* was proposed, where time is divided into a sequence of control phase and data (transmis-

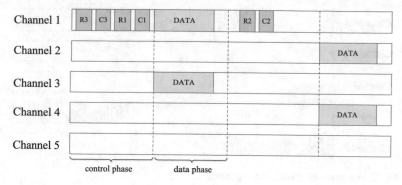

Fig. 3.5 Split phase method

sion) phase. All users utilize the dedicated control channel to make an agreement on the schedule of the channels during the control phase, and they communicate through the scheduled channels in the data transmission phase.

As depicted in Fig. 3.5, time is divided into a sequence of two phases (control phase and data phase) and the users can send RTS or CTS (denoted as R and C in the figure) messages during the control phase on the dedicated common control channel (channel 1 in the figure); then two pairs of users utilize different channels (channels 1 and 3 for the first data phase in the figure) for communication in the data transmission phase. Multichannel Access Protocol (MAP) [13] is one example of this method. Obviously, this method could increase the efficiency of using all channels, but the users need to be synchronized, i.e. they should start the process at the same time.

In view of the disadvantages surrounding the common control channel, many rendezvous protocols not based on a dedicated CCC were proposed and there are mainly two approaches: common hopping and parallel rendezvous [8, 35, 43, 45, 46].

The intuitive idea of *common hopping* is: all users hop through all channels synchronously. If a pair of devices (users) want to communicate with each other, they stop hopping and use the current channel. Once they finish, they rejoin the common hopping pattern just like the other users.

For example, there are five wireless channels as depicted in Fig. 3.6 and all users hop through the channels with the pattern 1, 2, 3, 4, 5, 1, 2, 3, 4, 5, ... If one device (called user A) wants to communicate with user B, it sends an RTS on the current hopping channel (for example, channel 5), and then user B responses with a CTS message on channel 5 if it is ready. Then both users stop hopping and use channel 5 for communication. Once the communication finishes, they rejoin the other users and continue hopping through the channels.

Channel Hopping Multiple Access (CHMA) [45] and Channel Hopping multiple Access with packet Trains (CHAT) [46] are two examples of this approach. The main advantage of this approach is that it uses all channels for communication,

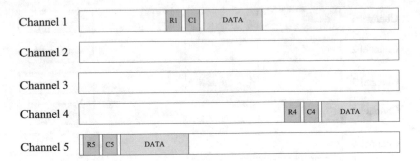

Fig. 3.6 Common hopping rendezvous protocol

which improves the efficiency of channels' usage. However, there are also some disadvantages:

(1) All users have to be synchronized, which means they have to start the process at the same time. This is a serious limitation in reality;
(2) once two users have occupied a channel for communication, the others may also still hop onto this channel. Clearly, this should be forbidden. Some works use carrier sensing to determine if the channel is busy, but this is costly in implementation [51].

The intuitive idea of *parallel rendezvous* is: the users can make an agreement on different channels by hopping through the channels according to different patterns. Slotted Seeded Channel Hopping (SSCH) [8] and Multi-channel MAC (McMAC) [43] are two examples of this method, where the users hop through the channels according to many different pseudorandom sequences that are decided by some seeds. The users can achieve rendezvous on different channels once the user is aware of the other user's seed and rendezvous happens on different channels.

3.5.3   Distributed Rendezvous Algorithms for Cognitive Radio Networks

However, these proposed rendezvous protocols are not applicable to cognitive radio networks (CRNs). There are two general types of rendezvous algorithms for CRN: centralized algorithms and decentralized algorithms.

Similar to the dedicated CCC method in traditional multichannel MAC protocols, *centralized rendezvous algorithms* for CRNs assume a central controller or the existence of a dedicated Common Control Channel (CCC) [1, 3, 32]. It is clear that the users can make an agreement through the central controller or the CCC, which simplifies the rendezvous process.

Although the centralized system is simple to implement and it can provide strong controllability of the channels, it is not flexible or scalable. There are several disadvantages.

(1) First, the central controller or the dedicated CCC can easily get congested as the number of users increases.

(2) Second, the centralized system is vulnerable to adversary attacks since the central controller or the CCC is vital for all rendezvous and for all the users in the network.

(3) Third, the cost to maintain the controller or the CCC is high because it needs to detect the channels' statuses, and to schedule the requests for communication from large quantities of wireless devices.

Therefore, failure or overloading of the central controller or the CCC could lead to dire consequences, and thus centralized algorithms are generally not practical.

Decentralized rendezvous algorithms without such central units have been proposed to avoid the drawbacks of centralized algorithms. There are mainly two categories depending on whether CCC is required.

Some decentralized algorithms establish local CCCs through which a user can contact their neighbors [2, 5]. These algorithms however incur substantial overhead in establishing and maintaining local CCCs.

Since central unit or the dedicated CCC has its inherent limitations, some researchers turned to decentralized algorithms without CCC, which are called *blind rendezvous algorithms* [1–4, 7, 8, 10–12, 17, 20, 26, 27, 36, 53]. Cognitive radios are autonomous and they are able to sense the licensed spectrum and access the available frequency bands on their own. Therefore, it is reasonable that the cognitive radios can communicate with others without relying on some other infrastructure such as some central unit or the CCC. Therefore, blind rendezvous algorithms are more practical and applicable for such intelligent cognitive radios, and large-scale CRNs can be constructed.

The existing blind rendezvous algorithms are divided into two classes: Global Sequence (GS) based rendezvous algorithms [7, 8, 21, 30, 40, 44] and Local Sequence (LS) based rendezvous algorithms [14, 15, 22, 24, 29].

The intuitive idea of *Global Sequence (GS)* based rendezvous algorithm is similar to that of the common hopping approach in traditional multichannel rendezvous protocols. According to the number of licensed channels and their labels, a sequence containing all these channels (labels of the channels) is constructed and all users (SUs) hop through the channels by repeating the sequence, which is called the *Global Sequence*.

For example, we can construct a sequence of length 15 on the basis of 3 channels $\{1, 2, 3\}$, as follows:

$$GS = \{1, 1, 2, 1, 3, 3, 3, 1, 2, 2, 3, 2, 1, 2, 3\} \tag{3.5}$$

Then two users can hop through the channels by repeating the sequence and they can achieve rendezvous at the same time no matter when they start. Suppose time is divided into slots of equal length and the user accesses a channel in each time slot

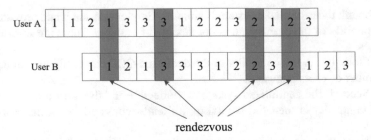

rendezvous

Fig. 3.7 An example of Global Sequence based rendezvous

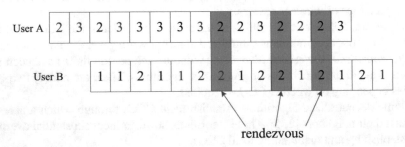

rendezvous

Fig. 3.8 An example of Global Sequence based rendezvous: replacement happens when some channel is not available

according to the sequence. As depicted in Fig. 3.7, two users can achieve rendezvous on all the three channels even if they are asynchronous, i.e. they do not start at the same time.

However, a user may not need to access all licensed channels according to the sequence since some of them may be occupied by the licensed users (PUs); then the user chooses another available channel randomly to continue. For example, in Fig. 3.8, channel 1 is not available for user A and it has to replace it by any available one from $\{2, 3\}$ (the channels in red represent the replaced ones). Similarly, channel 3 is not available for user B and so a replacement is also needed. Subsequently, the two users rendezvous on channel 2, as shown in the figure.

The first rendezvous happens since user A chooses a random available channel (2) to replace channel 1 in the sequence, and the other two rendezvous situations are generated according to the original channel hopping sequence.

Actually, if the constructed GS has some good properties, rendezvous can always be guaranteed once two users share at least one common available channel, no matter how many channels are available for each user and when do they start the process [3]. We call these GS based rendezvous algorithms and the technique to hop through the licensed channels according to some pre-defined sequences is called *Channel Hopping (CH)* [1, 2, 7, 8, 10, 11]. Jump-Stay (JS) algorithm [7], Channel Rendezvous Sequence (CRSEQ) [8] and Disjoint Relaxed Difference Set (DRDS) based rendezvous algorithm [21] are some state-of-the-art GS based rendezvous algorithms. We will introduce these algorithms in Chap. 8.

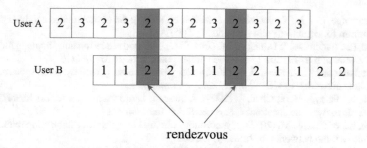

Fig. 3.9   An example of Local Sequence based rendezvous

The other type of blind rendezvous algorithm is *Local Sequence (LS)* based rendezvous algorithms, which is firstly defined in [4].

GS based rendezvous algorithms construct a fixed length sequence for all users and it is inefficient when the number of available channels accounts for a small fraction of all licensed channels. LS based algorithms can accelerate the rendezvous process and the intuitive idea is to construct different sequences for different users based on the local information.

In order to design different sequences, most works assume each user in the network has a distinct identifier (ID) and different sequences can be constructed on the basis of the user's ID and its sensed available channels. As depicted in Fig. 3.9, user *A* hops through its available channels {2, 3} by repeating the sequence {2, 3}, while user *B* accesses the available channels {1, 2} according to the sequence {1, 1, 2, 2}, rendezvous happens on channel 2 in a shorter time compared with Fig. 3.8.

The first algorithm belonging to this type is the Ring-Walk (RW) rendezvous algorithm [5]. Following that, Alternate Hop-and-Wait (AHW) [2], Local Sequence (LS) based rendezvous algorithm, and Modified Local Sequence (MLS) based rendezvous algorithm were proposed, where the the user's ID plays an important role in sequence construction. There are also some works that establish different sequences without the users' ID [1], which we will introduce later in this book.

References

1. Alpern, S. (1976). Hide and seek game. In *Seminar at the Institute für Höhere Studien*, Vienna.
2. Alpern, S. (1995). The rendezvous search problem. *SIAM Journal of Control and Optimization*, *33*(3).
3. Alpern, S., Baston, V. J., & Essegaier, S. (1999). Rendezvous search on a graph. *Journal of Applied Probability*, *36*(1), 223–231.
4. Alpern, S., & Beck, A. (1999). Asymmetric rendezvous on the line is a double linear search problem. *Mathematics of Operations Research*, *24*(3), 604–618.
5. Alpern, S., & Pikounis, M. (2000). The telephone coordination game. *Game Theory and Applications*, *5*, 1–10.
6. Anderson, E. J., & Weber, R. R. (1990). The rendezvous problem on discrete locations. *Journal of Applied Probability*, *28*, 839–851.

7. Arachchige, C. J. L., Venkatesan, S. & Mittal, N. (2008). An asynchronous neighbor discovery algorithm for cognitive radio networks. In *DySPAN*.
8. Bahl, P., Chandra, R., & Dunagan, J. (2004). SSCH: Slotted seeded channel hopping for capacity improvement in IEEE 802.11 Ad Hoc wireless networks. In *MobiCom*.
9. Balachandran, K., & Kang, J. H. (2006). Neighbor discovery with dynamic spectrum access in adhoc networks. In *Vehicular Technology Conference*.
10. Bian, K., Park, J.-M., & Chen, R. (2009). A quorum-based framework for establishing control channels in dynamic spectrum access networks. In *Mobicom*.
11. Bian, K., & Park, J.-M. (2011). asynchronous channel hopping for establishing rendezvous in cognitive radio networks. In *IEEE INFOCOM*.
12. Bian, K., Park, J.-M., & Chen, R. (2011). Control channel establishment in cognitive radio networks using channel hopping. *IEEE Journal on Selected Areas in Communications*, *29*(4), 689–703.
13. Chen, J., Sheu, S., & Yang, C. (2003). A new multichannel access protocol for IEEE 802.11 ad hoc wireless LANs. In *Proceeding of 14th IEEE International Symposium on Personal, Indoor and Mobile Radio Communications (PIMRC)*.
14. Chen, S., Russell, A., Samanta, A. & Sundaram, R. (2014). Deterministic blind rendezvous in cognitive radio networks. In *ICDCS*.
15. Chuang, I., Wu, H.-Y., Lee, K.-R., & Kuo, Y.-H. (2013). Alternate hop-and-wait channel rendezvous method for cognitive radio networks. In *INFOCOM*.
16. Conte, G., & Pennesi, P. (2010). The rendezvous problem with discontinuous control policies. *IEEE Transaction on Automatic Control*, *55*(1), 279–283.
17. Dai, Y., Wu, J., & Xin, C. (2013). Virtual Backbone Construction for Cognitive Radio Networks without Common Control Channel. In *INFOCOM*.
18. Dessmark, A., Fraigniaud, P., & Pelc, A. (2003). Deterministic rendezvous in graphs. In *ESA*.
19. Dudek, G., & Roy, N. (1997). Multi-robot rendezvous in unknown environments. In *AAAI*.
20. Gandhi, R., Wang, C.-C.. & Hu, Y.C. (2012). Fast rendezvous for multiple clients for cognitive radiso using coordinated channel hopping. In *SECON*.
21. Gu, Z., Hua, Q.-S., Wang, Y., & Lau, F. C. M. (2013). Nearly optimal asynchronous blind rendezvous algorithm for cognitive radio networks. In *SECON*.
22. Gu, Z., Hua, Q.-S., & Dai, W. (2014). Local sequence based rendezvous algorithms for cognitive radio networks. In *SECON*.
23. Gu, Z., Hua, Q.-S., Wang, Y. & Lau, F. C. M. (2014). Oblivious rendezvous in cognitive radio networks. In *SIROCCO*.
24. Gu, Z., Hua, Q.-S., & Dai, W. (2014). Fully distributed algorithms for blind rendezvous in cognitive radio networks. In *MOBIHOC*.
25. Hung, W.-C., Law, K. L. E., & Leon-Garcia, A. (2002). A dynamic multi-channel MAC for ad hoc LAN. In *Proceeding of 21st Biennial Symposium on Communications*.
26. Kondareddy, Y., Agrawal, P., & Sivalingam, K. (2008) Cognitive radio network setup without a common control channel. In *MILCOM*.
27. Lazos, L., Liu, S., & Krunz, M. (2009). Spectrum opportunity-based control channel assignment in cognitive radio networks. In *SECON*.
28. Lin, Z., Francis, B., & Maggiore, M. (2009). Getting mobile autonomous robots to rendezvous. *Control of uncertain systems*, pp. 119–137.
29. Lin, Z., Liu, H., Chu, X., & Leung, Y.-W. (2012). Ring-walk rendezvous algorithms for cognitive radio networks. *Ad Hoc & Sensor Wireless Networks*.
30. Liu, H., Lin, Z., Chu, X., & Leung, Y.-W. (2012). Jump-stay rendezvous algorithm for cognitive radio networks. *IEEE Transactions on Parallel and Distributed Systems*, *23*(10), 1867–1881.
31. Liu, H., Lin, Z., Chu, X. & Leung, Y.-W. (2012). Taxonomy and challenges of rendezvous algorithms in cognitive radio networks. In *ICNC*.
32. Lo, B. F. (2011). A survey of common control channel design in cognitive radio networks. *Physical Communication*, *4*, 26–39.
33. Marco, G. D., Gargano, L., Kranakis, E., & Krizanc, D. (2006). Asynchronous deterministic rendezvous in graphs. *Theoretical Computer Science*, *355*, 315–326.

34. Mo, J., So, H.-S. W., & Walrand, J. (2005). Comparison of multi-channel MAC protocols. In *MSWiM*.
35. Mo, J., So, H.-S. W., & Walrand, J. (2008). comparison of multichannel MAC protocols. *IEEE Transaction on Mobile Computing, 7*(1), 50–65.
36. Perez-Romero, J., Salient, O., Agusti, R., & Giupponi, L. (2007). A novel on-demand cognitive pilot channel enabling dynamic spectrum allocation. In *DySPAN*.
37. Romaszko, S. (2013). A rendezvous protocol with the heterogeneous spectrum availability analysis for cognitive radio ad hoc networks. *Journal of Electrical and Computer Engineering*.
38. Roy, N., & Dudek, G. (2001). Collaborative robot exploration and rendezvous: Algorithms, performance bounds and observations. *Autonomous Robots, 11*, 117–136.
39. Schenato, L., & Zampieri, S. (2006). Optimal rendezvous control for randomized communication topologies. In *CDC*.
40. Shin, J., Yang, D., & Kim, C. (2010). A channel rendezvous scheme for cognitive radio networks. *IEEE Communications Letters, 14*(10), 954–956.
41. Silvius, M. D., MacKenzie, A. B., & Bostian, C. W. (2007). A survey of neighbor discovery protocols in modern wireless systems and their application to the design and implementation of "rendezvous" algorithms in smart and cognitive radios. In *Virginia Tech Symposium on Wireless Personal Communications*.
42. So, J. & Vaidya, N. (2004). Multi-channel MAC for ad hoc networks: Handling multi-channel hidden terminals using a single transceiver. In *MobiHoc*.
43. So, H.W., Walrand, J., & Mo, J. (2007). McMAC: A multi-channel MAC proposal for ad hoc wireless networks. In *Proceeding of IEEE Wireless Communications and Networking Conference (WCNC)*.
44. Theis, N. C., Thomas, R. W., & DaSilva, L. A. (2011). Rendezvous for cognitive radios. *IEEE Transactions on Mobile Computing, 10*(2), 216–227.
45. Tzamaloukas, A., & Garcia-Luna-Aceves, J. J. (2000). Channel-hopping multiple access. In *ICC*.
46. Tzamaloukas, A., & Garcia-Luna-Aceves, J. (2000). Channel-hopping multiple access with packet trains for ad hoc networks. In *Proceeding of IEEE Device Multimedia Communications (MoMuC)*.
47. Wu, S.-L., Lin, Y., Tseng, Y.-C., & Sheu, J.-P. (2000). A new multi-channel MAC protocol with on-demand channel assignment for mobile ad hoc networks. Algorithms and Networks (ISPAN). In *Proceeding of International Symposium on Parallel Architectures*.
48. Wu, S.-L., Lin, C.-Y., Tseng, Y.-C., Lin, C.-Y., & Sheu, J.-P. (2002). A multi-channel MAC protocol with power control for multi-hop mobile ad hoc networks. *The Computer Journal, 45*(1), 101–110.
49. Wu, C.-C., & Wu, S.-H. (2013). On bridging the gap between homogeneous and heterogeneous rendezvous schemes for cognitive radios. In *MobiHoc*.
50. Wu, S.-H., & Wu, C.-C., Hon, W.-K., & Shin, K. G. (2014). Rendezvous for heterogeneous spectrum-agile devices. In *INFOCOM*.
51. Zannoth, M., Ruhlicke, T., & Klepser, B.-U. (2004). A highly integrated dual-band multimode wireless LAN transceiver. *IEEE Journal of Solid-State Circuits, 39*(7), 1191–1195.
52. Zhang, D., He, T., Ye, F., Ganti, R. & Lei, H. (2012). EQS: Neighbor discovery and rendezvous maintenance with extended quorum system for mobile sensing applications. In *ICDCS*.
53. Zhao, J., Zheng, H., & Yang, G.-H. (2005). Distributed coordination in dynamic spectrum allocation networks. In *DySPAN*.

Chapter 4
Rendezvous Categories

Abstract There are many kinds of rendezvous settings and we present their differences in this chapter. For simplicity, assume there are M entities (or users) in the distributed system, and each entity has N external ports (or channels) for rendezvous attempt. Due to environmental noise or the temporary port occupancy by others (such as other devices or services), some ports may not be usable for a long time. We differentiate between different rendezvous categories according to different settings. In Sect. 4.1, the entities may run the same or different algorithms, which are referred to as symmetric or asymmetric algorithms. In Sect. 4.2, we introduce synchronous and asynchronous settings, where the entities start the algorithms at the same time or at different times. In Sect. 4.3, the entities may have the same or different sets of available ports that are not occupied, and are called symmetric or asymmetric port setting. In Sect. 4.5, we introduce a special scenario called oblivious port labeling where the entities could label the ports locally. In contrast, all entities see the same labels of the ports in non-oblivious port labeling, which limits the power of a distributed system. Finally, we introduce different rendezvous categories in Sect. 4.6 according to different settings that are introduced above.

4.1 Symmetric and Asymmetric Algorithms

In a distributed system, all entities have the same role and they do not know the others' information or status. Therefore, all users (or entities) in the network should execute the same rendezvous algorithm, which is called *symmetric algorithm*. For example, a rendezvous strategy is to access each port round after round in a sequential order, and all users have to run the same strategy, which unfortunately may lead to infinite waiting for rendezvous.

Considering two users a and b, denote user a's labels of the ports as $\{a_1, a_2, \ldots, a_N\}$, and the ports' labels of user b as $\{b_1, b_2, \ldots, b_N\}$. If the ports between two users are connected as follows:

$$port\ a_i \leftrightarrow port\ b_i, 1 \leq i \leq N \tag{4.1}$$

© Springer Nature Singapore Pte Ltd. 2017
Z. Gu et al., *Rendezvous in Distributed Systems*,
DOI 10.1007/978-981-10-3680-4_4

They will choose the connected ports a_i, b_i all the time as time goes on. However, suppose the ports between two users are connected as follows:

$$\begin{cases} port\ a_i \leftrightarrow port\ b_{i+1},\ 1 \le i \le N - 1 \\ port\ a_N \leftrightarrow port\ b_1 \end{cases} \tag{4.2}$$

Suppose the two users run the same algorithm by accessing the ports sequentially at the same time; they will choose the ports with same label a_i, b_i all the time, and rendezvous can never happen.

In some rendezvous settings, the users are allowed to run different algorithms, which are called *asymmetric algorithms*. This is because there are two types of roles in the users' communication: transmitter who sends the data to the neighbor, and receiver who receives the message. In addition, some users may dominate their neighbors and they could have different roles. Therefore, some works assume the users can perform asymmetric algorithms to guarantee rendezvous.

Considering the above example where port a_i is connected to port b_{i+1}, if two users are able to run asymmetric algorithms, rendezvous can be guaranteed within finite time.

Suppose user a chooses a fixed port a_1 (for example), while user b adheres to the above strategy, by choosing the ports sequentially, i.e. $\{b_1, b_2, \ldots, b_N, b_1, b_2, \ldots, b_N\}$. Clearly, two users can achieve rendezvous when user b chooses the port b_2.

Nevertheless, asymmetric algorithms are not preferable in distributed systems, because it is hard to assign different roles to or distinguish among the entities when there are a large number of entities in the network. Therefore, many works have focused on symmetric algorithm design and this is also our focus in this book.

4.2 Synchronous and Asynchronous

Timing is a crucial element in distributed computing. In synchronous settings, all users in the system can access a global clock and the time can be considered as divided into slots of equal length, and all users start the distributed algorithms from the same time slot. This assumption simplifies the rendezvous problem since we do not need to distinguish between the two users as one user may otherwise start the rendezvous search while the other is still asleep.

In asynchronous settings, any user in the network can start the rendezvous process at any time, which is more practical since the entities in a distributed system are not usually aware of the others' information, including what the other entities do and when they start. Therefore, most works study asynchronous settings in distributed computing, including the rendezvous problem. The challenge of designing asynchronous algorithms is: a user in the network may start the algorithm at any global time, and the asynchronous algorithm has to be robust enough so that all users can finish their computational tasks efficiently regardless of when they start the algorithm.

We define these two settings in a more formal way. Suppose there are M users in the distributed system, which for simplicity we denote as

$$\{u_1, u_2, \ldots, u_M\}.$$

Actually, the users in a distributed system may not have such identifiers, but we use the notation to distinguish them from a global view. Suppose the users are to solve a computational problem P and the proposed algorithm is denoted as F (if the users choose a symmetric algorithm as discussed in the above section; we would denote the adopted algorithms for each user as $\{F_1, F_2, \ldots, F_M\}$ in the asymmetric case).

Suppose there is a global clock which records the detailed time, but the users may not be able to access the clock. Denote the start time of the users according to the global clock as

$$\{t_1, t_2, \ldots, t_M\}.$$

However, for each user in the network, its local time is just 0 when it starts computing the problem.

In synchronous setting, all users start the computation at the same time, which implies:

$$t_1 = t_2 = \ldots = t_M. \tag{4.3}$$

The users may finish the computational task with different times according to their own resources and local memory. It is easy to define when the distributed system finishes solving the problem, i.e. the time at which the last user finishes computing.

In asynchronous setting, the users may start the computation with no time reference, which may result in different start-up times from the global clock view, i.e. t_i could be equal to, smaller than or larger than t_j for any $i, j \in [1, M]$. Similar to synchronous setting, the users may spend different times to finish their computations and it is harder to define the time complexity of solving a problem in such a distributed system.

Suppose the finish times of all users are

$$\{d_1, d_2, \ldots, d_M\}$$

respectively, where $t_i < d_i, \forall i \in [1, M]$. A simple way of defining the rendezvous time complexity is to compute the elapsed time from when the first user starts the algorithm to the time when the last user finishes; the time complexity is defined as:

$$Complexity = \max_{1 \le i \le M} d_i - \min_{1 \le i \le M} t_i. \tag{4.4}$$

However, this definition may not be reasonable, since some user in the distributed system may start the computation very late and it will result in a very large time computational complexity. Moreover, the message delivered in the system may have unbounded time, which makes the time complexity hard to compute. In order to

Fig. 4.1 An example of
computing time complexity
in a distributed system

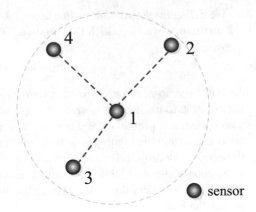

eliminate the effect of message delay, many works assume that the message delay
has a finite bound.

One reasonable way is to compute the time cost by each user and the largest cost
among all users' costs is the regarded as the time of the asynchronous algorithm in
the distributed system. We define the time complexity as:

$$Complexity = \max_{1 \leq i \leq M} (d_i - t_i). \tag{4.5}$$

However, this method may not work well under some conditions, especially when
one user's computation is based on another's information.

For example, in Fig. 4.1, suppose each sensor detects a value of the environment
and all sensors have to compute the local maximum value among all neighbors.
Assume sensor 1 starts at a very early time, while its neighbor sensor 2 starts very
late. Even after sensor 1 receives messages from both sensors 3 and 4, it cannot
terminate until sensor 2 starts the algorithm by sending the value to it. Therefore, the
delay of sensor 2 increases the time complexity of the algorithm significantly.

There is another way to compute the time complexity of an asynchronous algorithm. Since the distributed system is composed of all M users, and we define the
start time of the system as the time when the last user starts the computation, i.e.

$$t_{DS} = \max_{1 \leq i \leq M} t_i \tag{4.6}$$

is the system start time. Using a similar idea, we define the finish time of the distributed system as the time when all users have finished their computation, i.e.

$$d_{DS} = \max_{1 \leq i \leq M} d_i \tag{4.7}$$

is the system finish time.

Then, we define the time complexity that the distributed system finishes the computation as the elapsed time between the system start time and the system finish time, i.e.

$$Complexity = d_{DS} - t_{DS} = \max_{1 \leq i \leq M} d_i - \max_{1 \leq i \leq M} t_i.$$

This definition computes the elapsed time with the assumption that the network is a unified system. We only consider when all users have started the computation and when all users have finished the computation. The definition could resolve the ambiguity that some sole user in the system may affect and increase the computation time. In this book, we will adopt this definition when computing time complexity to rendezvous in asynchronous settings.

4.3 Symmetric and Asymmetric Port Settings

In a cognitive radio network, although there may be N licensed channels that can be detected and sensed by the secondary users (such as mobile phones, computers, etc.), some licensed channels may be occupied by nearby licensed users who paid and owned the licensed channels. Due to the different occupancies of the licensed channels by different licensed users, different secondary users could have different *available* licensed channels for further use. Recall Fig. 3.3 as an example, user A can sense three available licensed channels $\{1, 2, 5\}$ but user B can sense three available channels $\{3, 4, 5\}$. Then, two users have to attempt to find the common available channel 5 through their designed distributed algorithms. Hence, the secondary users may find different sets of available channels, which is called an *asymmetric* situation.

Similarly, suppose each user has N external ports for connection, but not all of them are available for a certain specific user. As shown in Fig. 4.2, node a and node b are two neighbors and they may try to communicate with each other by constructing

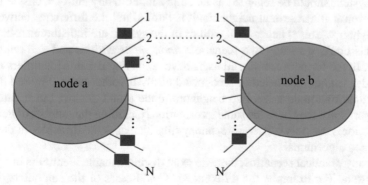

Fig. 4.2 An example of port occupancy in rendezvous settings

a connection link. However, some external ports of node a may be used for some other services, or some ports are not available for data transmission. The ports that have already been occupied are blocked with a red rectangle in the figure.

Therefore, node a may get a set of available ports as set P_a. Similarly, node b can get a set of available ports as set P_b. For simplicity, we assume the nodes have the same labels for N external ports, and port i of node a is connected to port i of node b. If two nodes share the same set of available ports, rendezvous is easier to achieve and we call this situation the *symmetric port setting*. Alternatively, in the *asymmetric port setting*, $P_a \neq P_b$ between two nodes. This makes rendezvous more difficult because if node a accesses some channel i in set P_a but not in set P_b, there is no chance for the two users to rendezvous, no matter which port node b chooses. For example, if node a uses port 4 in the figure, but it is blocked for user b, then the two users can never rendezvous for communication if node a does not change ports.

Symmetric setting is a special case of port setting and it can verify how the designed algorithms perform with a relatively "good" setting. Some works also study a more restricted setting, where all ports are available, i.e. $P_i = \{1, 2, \ldots, N\}$ and this can be used to verify the "good part" of any distributed rendezvous algorithm.

However, symmetric setting is rare in practical due to uncontrollable environmental noise and undecidable port occupancy. Most works study the asymmetric setting and design distributed algorithms that work efficiently under both symmetric and asymmetric port settings. There exists an extreme situation in asymmetric setting, where two nodes have only 1 common available port, i.e. $|P_1 \bigcap P_2| = 1$; there is still chance to rendezvous, but it is the most difficult situation to handle for any distributed rendezvous algorithm. This extreme setting can be used to test the efficiency of any distributed rendezvous algorithm in the worst situation.

4.4 Anonymous and Non-anonymous Entities

In a distributed system, the entities are distributed with no pre-defined rules and the whole system should be regarded as an empty input to any entity. Considering any entity, even if it has several neighbors, it is hard to tell the difference between any two neighbors. This is because all entities in the system are indistinguishable.

In this book, we study two scenarios: *anonymous entities* and *non-anonymous entities*. In the former scenario, all users have no distinguishable identifiers and any two nodes cannot be labeled or recognized easily by other users. Figure 4.3 shows an example that node a has two neighbors, node b and node c, but it cannot tell them apart since they have the same information (including the number of neighbors or some local memory). Therefore, anonymity increases the difficulty in designing distributed algorithms.

In many practical scenarios, there do exist distinguishable identifiers in a distributed system. For example, the different MAC addresses of the computers, the IDs of sensor nodes, etc. Therefore, the users in Fig. 4.4 can be distinguished by their identifiers (IDs) and that makes many problems easier to solve.

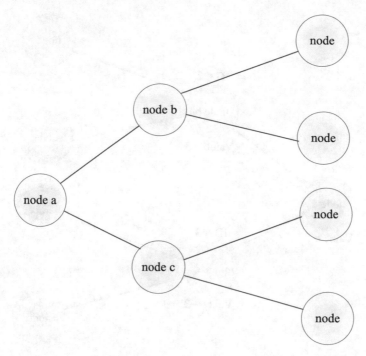

Fig. 4.3 An example of anonymous network where node *a* cannot distinguish node *b* and node *c*

For example, if node *a* wants to collect data from its neighbors in Fig. 4.4. If the two neighbors are anonymous as shown in the figure, after receiving two exact values 3, node *a* cannot decide whether to terminate or to wait for more data, since these two values may come from the same neighboring node *b* or node *c* (we assume node *b* and node *c* will both send their value continuously until node *a* terminates). However, if the nodes are non-anonymous, node *a* could use the identifier of each sender to decide whether two neighbors have already sent their values.

In a rendezvous problem, users' identifiers may not play an important role in designing efficient algorithm, but they do matter in certain types of rendezvous. We will introduce rendezvous for both the anonymous and the non-anonymous settings.

4.5 Oblivious and Non-oblivious Port Labeling

In wireless networks, channels are always labeled by some pre-defined rules. Therefore, all the channels (spectrum) share the same global labeling for all users in the network. However, when two entities try to construct a communication link through their external ports in a distributed system, these ports may not share the same labels, and insisting on connecting ports with the same label will amount to trouble. Therefore, we need to address the significance of port labeling.

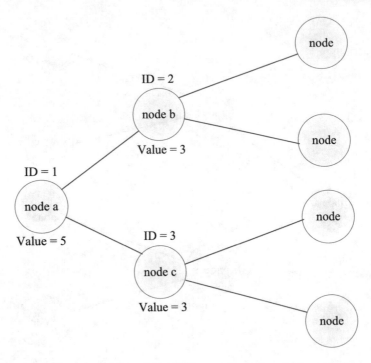

Fig. 4.4 An example of non-anonymous network

In some distributed systems, for example Cognitive Radio Networks (CRNs), the secondary users can communicate with others through available licensed channels. These licensed channels correspond to the external ports in our studied rendezvous problem. Moreover, these licensed channels are always assumed to be assigned with identifiers (IDs) and the channel IDs are known to all secondary users. This assumption makes rendezvous easier since two secondary users can design rendezvous algorithms based on the channel IDs. Considering the simple algorithm:

* *The user chooses channel $(i - 1)\%N + 1$ in the i-round.*

Here N represents the number of all channels and the users make decisions round by round.

In the simplest version, two users start the algorithm at the same time, and they will always choose the same channel in each round. Even if some channels are not accessible to secondary users due to the licensed users' occupancy, they can achieve rendezvous definitely if they share some common available channels. Moreover, the time to rendezvous is very small (no larger than N rounds).

However, a global labeling could not simplify the rendezvous problem much. We can use the above simple algorithm as the example again. If two users are asynchronous, i.e. they start the rendezvous attempt in different rounds, they may never rendezvous even all channels are available for both users. This is easy to check, since when one user chooses channel i, the other one will choose channel $(i + \Delta - 1)\%N + 1$ where Δ means how many rounds the other one is late by. Obviously, if $\Delta \% N \neq 0$, these two channels have different labels, and they can never rendezvous.

In a distributed system, it is hard to make all users share the same labels. For example, in a cognitive radio network, the 'TV white space' that could be sensed by the users has operating frequencies ranging from 470–790 MHz in Europe [1, 3], but it is located in the VHF (i.e. very high frequency) (54–216 MHz) and UHF (i.e. ultra high frequency) (470–698 MHz) bands in the United States [2]. Obviously, the labeling of this space could vary and the same frequency band (channel) may be assigned different labels under different administrations.

In studying a more general rendezvous problem in a distributed system, we may need to observe the fact that the external ports are oblivious, which means they have no global labels and the other nodes cannot tell the difference of these ports. This assumption makes us dig deep into the core of rendezvous in distributed systems and the solutions then derived can be applied to many interesting applications.

4.6 Rendezvous Categories

According to the five dimensions of rendezvous described earlier, we denote a specific rendezvous setting by a five element array:

$$RS = <Alg, Time, Port, ID, Label> \tag{4.8}$$

where each element has two options:

(1) Alg: *Asymmetric* or *Symmetric* algorithms as in Sect. 4.1 (we use *Alg-AS* and *Alg-S* for short);
(2) Time: *Synchronous* or *Asynchronous* system as in Sect. 4.2 (we use *Syn* and *Asyn* for short);
(3) Port: *Symmetric* or *Asymmetric* of available ports as in Sect. 4.3 (we use *Port-S* and *Port-AS* for short);
(4) ID: *Non-Anonymous* or *Anonymous* entities as in Sect. 4.4 (we use *Non-Anon* and *Anon* for short);
(5) Label: *Non-oblivious* or *Oblivious* port labeling as in Sect. 4.5 (we use *Non-Obli* and *Obli* for short).

Clearly, there are $2^5 = 32$ different types of rendezvous settings which we will introduce in this book. Notice that, the easiest rendezvous setting should be

$$RS_e = <Alg - AS, Syn, Port - S, Non - Anon, Non - Obli> \tag{4.9}$$

while the most difficult setting should be

$$RS_h = <Alg - S, Asyn, Port - AS, Anon, Obli> \tag{4.10}$$

For two different settings RS_i, RS_j, we define $RS_i \preceq RS_j$ if RS_i is not more difficult than RS_j. We introduce the following property about the hardness of different rendezvous settings.

Proposition 4.1 *Considering two settings:*

$$RS_a = < Alg - AS, Time, Port, ID, Label >$$

$$RS_b = < Alg - S, Time, Port, ID, Label >$$

we have $RS_a \preceq RS_b$.

For any fixed setting of $Time, Port, ID, Label$, designing an asymmetric rendezvous algorithm is not more difficult than designing a symmetric algorithm. This is because the entities could also adopt the same algorithm in asymmetric algorithm setting. Actually, for the distributed rendezvous problem, the entities running symmetric algorithms are much harder to achieve rendezvous than running asymmetric algorithms. We will show some examples in the following chapters.

Proposition 4.2 *Considering two settings:*

$$RS_a = < Alg, Syn, Port, ID, Label >$$

$$RS_b = < Alg, Asyn, Port, ID, Label >$$

we have $RS_a \preceq RS_b$.

For any fixed setting of $Alg, Port, ID, Label$, designing an asynchronous algorithm is not easier than designing a synchronous algorithm. This is because asynchronous setting contains the situation of synchronous setting when the entities start at the same time. In some settings, designing efficient synchronous algorithms could be as hard as designing efficient asynchronous algorithms, and we will introduce some of them in Part III.

Proposition 4.3 *Considering two settings:*

$$RS_a = < Alg, Time, Port - S, ID, Label >$$

$$RS_b = < Alg, Time, Port - AS, ID, Label >$$

we have $RS_a \preceq RS_b$.

For any fixed setting of *Alg, Time, ID, Label*, it is not easier to handle when the ports are asymmetric than the symmetric port situations. When the entities have symmetric ports, each attempted port for rendezvous could match the other port, but it may not happen when the ports are asymmetric.

Proposition 4.4 *Considering two settings:*

$$RS_a = < Alg, Time, Port, Non - Anon, Label >$$

$$RS_b = < Alg, Time, Port, Anon, Label >$$

we have $RS_a \preceq RS_b$.

For any fixed setting of *Alg, Time, Port, Label*, the entities with distinguishable identifiers could achieve rendezvous easier or not more difficult than when they are anonymous. This is obvious but under most conditions, entities' IDs do not play a vital role in achieving rendezvous.

Proposition 4.5 *Considering two settings:*

$$RS_a = < Alg, Time, Port, ID, Non - Obli >$$

$$RS_b = < Alg, Time, Port, ID, Obli >$$

we have $RS_a \preceq RS_b$.

For any fixed setting of *Alg, Time, Port, ID*, if the ports are labeled with some pre-defined rules, the rendezvous problem would become much easier. This is because many rendezvous algorithms utilize the ports' labels to design hopping sequences and the global information could help guarantee rendezvous. If the ports have no common labels, the entities could do less to walk out the difficult position, even if they could do labeling on their own.

Among the five fundamental elements of any rendezvous setting, the labeling of ports plays the most important role and there is a big difference when the ports are oblivious. In the chapters that follow, we divide and treat rendezvous algorithms in two parts on the basis of port labeling settings.

In Part II, we assume all entities have the same labels of all the ports, which is also the assumption by most extant works in multichannel wireless networks, cognitive radio networks, etc. We present different types of rendezvous algorithms for the $2^4 = 16$ settings when the ports are non-oblivious. In Part III, all ports are assumed to be oblivious, i.e. there is no common or global labeling of the ports, and we introduce rendezvous algorithms for the other $2^4 = 16$ settings.

References

1. ETSI. (2012). EN 301 598 White Space Devices (WSD); Wireless Access Systems Operating in the 470 MHz to 790 MHz Frequency Band.
2. Flores, A. B., Guerra, R. E., & Kightly, E. W. (2013). IEEE 802.11af: A standard for TV white space spectrum sharing. *IEEE Communications Magazine*.
3. Ofcom. (2013). Regulatory requirements for white space devices in the UHF TV band. http://www.cept.org/Documents/se-43/6161/.

Part II
Blind Rendezvous in Distributed Systems

Chapter 5
Blind Rendezvous Problem

Abstract Rendezvous is the fundamental process to establish a communication link between a pair of neighboring entities. In traditional multichannel wireless networks and cognitive radio networks, rendezvous is the prerequisite for communication, via which the users try to choose the same channel for data transmission. Here, we introduce the *blind rendezvous problem*, where *blind* means the entities or the users in the system do not know the others' information and they have to make decisions completely locally. This definition makes a distinction away from *centralized rendezvous* where a central unit is used to provide the port or channel information to the users [1, 3], or some local common control channel is established and maintained to control and simplify the rendezvous process [2, 5]. Blind rendezvous draws a lot of attention from both academic and industrial areas due to its scalability, flexibility and robustness in implementing large scale distributed systems. We depict the blind rendezvous problem in Figs. 5.1 and 5.2. Consider a cognitive radio network which is composed of several secondary users (SUs) and several primary users (PUs). Because of the PUs' occupancy on the licensed channels, the SUs can only have opportunistically a portion of the licensed spectrum. Suppose user A has three channels $\{1, 2, 6\}$ that are not used by the PUs, while user B can access channels $\{3, 5, 6\}$ after spectrum sensing. If they try to communicate with each other, they should choose an available channel for their communication attempt. However, neither of them knows the other SU's information about the licensed channels, so they have to apply rendezvous strategies in a distributed "blind" way. Consider a simple algorithm: each SU accesses the available licensed channels in a round robin way, i.e. user A accesses channels by repeating the sequence $\{1, 2, 6\}$:

$$\{1, 2, 6, 1, 2, 6, 1, 2, 6, 1, 2, 6, \ldots\} \tag{5.1}$$

and user B accesses channels by repeating the sequence $\{3, 5, 6\}$:

$$\{3, 5, 6, 3, 5, 6, 3, 5, 6, 3, 5, 6, \ldots\} \tag{5.2}$$

© Springer Nature Singapore Pte Ltd. 2017
Z. Gu et al., *Rendezvous in Distributed Systems*,
DOI 10.1007/978-981-10-3680-4_5

Time	1	2	3	4	5	6	7	8	9	10	11	...
User A	1	2	6	1	2	6	1	2	6	1	2	...
User B	3	5	6	3	5	6	3	5	6	3	5	...

Fig. 5.1 Blind rendezvous example for two synchronous users

Time	1	2	3	4	5	6	7	8	9	10	11	...
User A	1	2	6	1	2	6	1	2	6	1	2	...
User B		3	5	6	3	5	6	3	5	6	...	

Fig. 5.2 Blind rendezvous example for two asynchronous users

As shown in Fig. 5.1, users A and B can rendezvous on channel 5 at time 3 if they start at the same time, which is the *synchronous* setting. However, if user A is two time slots earlier than user B, which is the *asynchronous* setting, as shown in Fig. 5.2, they will never rendezvous. In this chapter, we first present the system model of blind rendezvous in Sect. 5.1, and then we introduce two important metrics: expected time to rendezvous and maximum time to rendezvous in Sect. 5.2. The problem definition is given in Sect. 5.3 and the challenges are discussed in Sect. 5.4. Finally, Sect. 5.5 summarizes the chapter.

5.1   System Model

In this part, we present the rendezvous algorithms for different types of rendezvous settings, on the basis that the ports are non-oblivious, i.e. all entities apply the same labeling rules. Therefore, the rendezvous settings can be represented as:

$$RS_{blind} = < Alg, Time, Port, ID, Non - Obli > \qquad (5.3)$$

where $Alg \in \{Alg-AS, Alg-S\}$, $Time \in \{Syn, Asyn\}$, $Port \in \{Port-S, Port-AS\}$, and $ID \in \{Non - Anon, Anon\}$.

We suppose there are $M(M \geq 2)$ users in a distributed system, and each user has $N(N \geq 1)$ external ports. In the RS_{blind} setting, the ports have the same labels from $\{1, 2, \ldots, N\}$ and for any two neighboring users, the ports with the same label are connected.

Fig. 5.3 Transform non-aligned slots to aligned ones

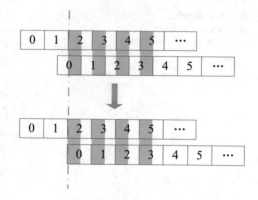

Denote all users as $\{u_1, u_2, \ldots, u_M\}$ and suppose their adopted rendezvous algorithms are $\{F_1, F_2, \ldots, F_M\}$ respectively (i.e. user u_i runs algorithm F_i).

(1) In $Alg - AS$ setting, for any two users $u_i, u_j, i \neq j$, F_i and F_j could be different (we use $F_i \neq F_j$ to denote that they are different);
(2) In $Alg - S$ setting, all users run the same algorithm, i.e. $\forall i, j \in [1, M]$, $F_i = F_j$.

Assume time is divided into slots of equal length $2t$, where t is sufficient for establishing a communication link between two users if they access the ports with the same label. In wireless networks, according to IEEE 802.22 [4], $t = 10$ ms and thus each time slot has a duration of 20 ms. Actually, the system may not be slot-aligned, which means the time clock of two users are not aligned. We show the method of transforming these situations to the slot-aligned scenario. For example, in Fig. 5.3, slot $i + 2$ of the upper row intersects slot i of the lower row and the intersection length is no less than t. Therefore, we transform it to the slot-aligned scenario since an overlap of t time length exists for establishing a communication link if they achieve rendezvous. In Fig. 5.4, we present another situation that the slots may not be aligned and the transformation is easy to see. Therefore, the slot length $2t$ is essential in designing rendezvous algorithms for asynchronous users and we can then consider the system as slot-aligned with this setting.

Denote the start time of the users as $\{t_1, t_2, \ldots, t_M\}$ respectively (i.e. the start time of user u_i is t_i).

(1) In Syn setting, all users have the same start time, i.e. $\forall i, j \in [1, M]$, $t_i = t_j$;
(2) In $Asyn$ setting, for any two users $u_i, u_j, i \neq j$, t_i and t_j could be different, i.e. $t_i \neq t_j$.

Due to occupancy by other services, each user could use only a portion of the external ports. We say a port is *available* if it is not occupied by other services and the user can choose it for communication. Denote the set of all external ports as $U = \{1, 2, \ldots, N\}$. For any user u_i, denote the set of available ports as $C_i \subseteq U$.

(1) In $Port - S$ setting, all users have the same available ports, i.e. $\forall i, j \in [1, M]$, $C_i = C_j$;

Fig. 5.4 Another
transformation from
non-aligned slots to aligned
ones

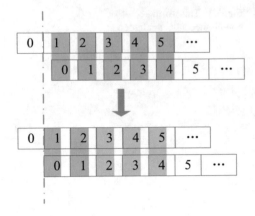

(2) In *Port − AS* setting, for any two users u_i, u_j, $i \neq j$, C_i and C_j could be different,
   i.e. $C_i \neq C_j$.

   In the *Port − AS* setting, in order to guarantee rendezvous, two neighboring users
must have at least one common available port, i.e. $C_i \bigcap C_j \neq \emptyset$.

   Considering all users in the system, some works assume they have distinct identifiers (IDs). We define the two settings as follows:

(1) In *Anon* setting, all users are anonymous and they have no distinct identifiers;
(2) In *Non − Anon* setting, each user has a distinct identifier (ID). Denote user u_i's
   ID as I_i. For any two users u_i, u_j, $i \neq j$, I_i and I_j are different, i.e. $I_i \neq I_j$.

5.2   Metrics

We use *Time to Rendezvous* (*TTR*) to measure the efficiency of rendezvous algorithms. As introduced in the model, the start time of user u_i is denoted as t_i. Suppose
the finish time of user u_i is d_i, where $d_i > t_i$. Notice that finish time means the
user finishes achieving rendezvous. For any two neighboring users u_i and u_j, if they
achieve rendezvous, both users will finish the process at the moment of rendezvous,
and thus $d_i = d_j$.

   We define the time to rendezvous between two users as follows.

Definition 5.1 For two neighboring users u_i and u_j, suppose their start times are
t_i, t_j respectively, and their finish times are d_i, d_j where $d_i = d_j = d$. The time to
rendezvous is defined as:

$$TTR = d - \max\{t_i, t_j\} \tag{5.4}$$

   We denote the rendezvous time as the time the user who starts later spends in the
process. Then, we define the time to rendezvous among all *M* users as follows.

Definition 5.2 Considering all users $u_1, u_2, \ldots, u_M$ in the system, denote their start times and finish times as $t_1, t_2, \ldots, t_M$ and $d_1, d_2, \ldots, d_M$ respectively. The time to rendezvous is defined as:

$$TTR = \max_{1 \le i \le M} d_i - \max_{1 \le i \le M} t_i \tag{5.5}$$

Since we consider different settings of rendezvous, different initial configurations (such as different number of available ports and different IDs) may lead to different times to rendezvous. Therefore, we use two metrics for evaluation:

(1) *Maximum Time to Rendezvous (MTTR)* represents the maximum time used to rendezvous among all different configurations;
(2) *Expected Time to Rendezvous (ETTR)* represents the expected time used to rendezvous among all different configurations.

MTTR reveals the performance of the rendezvous algorithm under the worst situation, while *ETTR* indicates how the algorithm performs on average.

5.3 Problem Definition

As described in the System Model, there are M users and their available ports are $C_1, C_2, \ldots, C_M$ respectively. If rendezvous happens for all users, denote the common available port set as $G = \bigcap_{i=1}^{M} C_i \ne \emptyset$. Before we define the rendezvous problem for multiple users in the system, we first formulate the rendezvous problem between two users.

Considering two users u_i and u_j, the set of available ports are C_i, C_j, and the IDs are I_i, I_j respectively. We formulate the *Blind Rendezvous Problem between Two Users* as follows.

Problem 5.1 Given a set $C_i \subseteq U$, design a port access algorithm f_i for different time slots $f_i(t) \in C_i$ such that: $\forall C_i, C_j \subseteq U, \forall \delta_t$:

$$\exists T \text{ s.t. } f_i(T + \delta_t) = f_j(T) \tag{5.6}$$

The *MTTR* value of algorithms f_i, f_j is $MTTR_{f_i, f_j} = \max_{\forall \delta_t} T$. The objective is to find algorithms minimizing the *MTTR* value among all algorithms: $MTTR = \min_{\forall f_i, f_j} MTTR_f$.

Remark 5.1 If user A starts later than user B, $\delta_t < 0$ in the description of Problem 5.1.

Notice that, if we are aiming at *symmetric algorithms*, both users u_i and u_j should adopt the same algorithm, i.e. $f_i = f_j$.

For example, $U = \{1, 2, 3, 4, 5, 6, 7, 8\}$, $C_i = \{1, 2, 5\}$, $C_j = \{3, 4, 5, 6\}$, and user u_i accesses the ports at different time slots, as follows:

Time	0	1	2	3	4	5	6	7	8	9	10	11	12	13	14	15
User u_i	1	2	5	1	2	5	1	2	5	1	2	5	1	2	5	1
User u_j	3	3	3	3	4	4	4	4	5	5	5	5	6	6	6	6

Fig. 5.5 An example of rendezvous between two synchronous users: users u_i and u_j start at time 0 simultaneously; and rendezvous can be achieved at time 8

Time	0	1	2	3	4	5	6	7	8	9	10	11	12	13	14	15
User u_i	-	-	-	-	1	2	5	1	2	5	1	2	5	1	2	5
User u_j	3	3	3	3	4	4	4	4	5	5	5	5	6	6	6	6

Fig. 5.6 An example of rendezvous between two asynchronous users: user u_j starts at time 0 while user u_i starts at time 4; then rendezvous is achieved at time 6

$$f_i(t) = \begin{cases} 1 \text{ When } t \equiv 0 \mod 3 \\ 2 \text{ When } t \equiv 1 \mod 3 \\ 5 \text{ When } t \equiv 2 \mod 3 \end{cases} \tag{5.7}$$

whereas user u_j accesses the ports by repeating the sequence:

$$\{3, 3, 3, 3, 4, 4, 4, 4, 5, 5, 5, 5, 6, 6, 6, 6\} \tag{5.8}$$

If they start at the same time, $TTR = 9$ when they both access port 5, as shown in Fig. 5.5. However, $TTR = 6$ when user u_j starts the process at time slot 0 but user u_i starts at time slot 4, as shown in Fig. 5.6.

In the example, two users adopt different algorithms and they are in the $Alg - AS$ setting. But it is impractical to design different strategies for different users, especially where the number of users M is large. Therefore, generally, symmetric algorithms for the rendezvous problem are preferred. In the example, $C_i \neq C_j$ is the $Port-AS$ setting and it would be the $Port - S$ setting if $C_i = C_j$. Both scenarios should be considered in order to guarantee rendezvous. Figure 5.5 shows the rendezvous between two synchronous users while Fig. 5.6 shows the rendezvous between two asynchronous users. Both scenarios should also be considered to guarantee rendezvous.

Based on the problem definition for two users, we formulate the *Blind Rendezvous Problem for Multiple Users in the Multihop system* as follows.

Problem 5.2 Considering M users $u_1, u_2, \ldots, u_M$, suppose their sets of available ports are $C_1, C_2, \ldots, C_M$, their IDs are $I_1, I_2, \ldots, I_M$, and their start times are

$t_1, t_2, \ldots, t_M$ respectively. Assume the diameter of the system is D, i.e. the minimum number of hops between any two users is no larger than D. For any set $C_i \subseteq U$, design a channel access algorithm for different time slots $f_i(t) \in C_i$ such that: for any two neighboring users $u_i, u_j, \forall \delta_t$:

$$\exists T \text{ s.t. } f_i(T + \delta_t) = f_j(T) \tag{5.9}$$

Each user u_i adopting algorithm f_i will finish the algorithm after achieving rendezvous with all neighbors by choosing a common port in $G = \bigcap_{j=1}^{M} C_j$ as the others. Suppose the finish time of user u_i is d_i. The time to rendezvous for the system is defined as:

$$TTR_{\delta_t} = \max_{i \in [1,M]} d_i - \max_{i \in [1,M]} t_i \tag{5.10}$$

The *MTTR* value of these algorithms is defined as:

$$MTTR_{f_1, f_2, \ldots, f_M} = \max_{\forall \delta_t} TTR_{\delta_t} \tag{5.11}$$

The objective is to find algorithms $f_1, f_2, \ldots, f_M$ minimizing the *MTTR* value among all strategies.

5.4 Challenges

In designing efficient distributed rendezvous algorithms, the most difficult rendezvous setting should be:

$$RS = < Alg - S, Asyn, Port - As, Anon, Non - Obli > \tag{5.12}$$

where the ports are defined to be non-oblivious, the rendezvous algorithms should be *symmetric* for the *asynchronous* users, with *asymmetric* port situations, and the users are *anonymous*.

In most situations, the users are not aware of which type of rendezvous setting they are in, and the designed general algorithm should work for all 16 different settings. We face the following challenges:

(1) Blind rendezvous algorithms should be efficient for both *port-symmetric* and *port-asymmetric* users;
(2) blind rendezvous algorithms should be applicable to both *synchronous* and *asynchronous* users;
(3) blind rendezvous algorithms prefer not using the users' IDs as input;
(4) blind rendezvous algorithms prefer symmetric algorithms for all users.

5.5   Chapter Summary

In this chapter, we introduce the blind rendezvous problem (BRP) in distributed systems. We describe the system model of blind rendezvous when the labels of the ports are non-oblivious, i.e. the external ports are labeled by a universal labeling rule. Then, we introduce two commonly used metrics for evaluating the rendezvous algorithms: expected time to rendezvous (*ETTR*) and maximum time to rendezvous (*MTTR*), which measure the average and worst case performance of the proposed algorithms respectively. We also give simple examples to illustrate the rendezvous between two users and list the challenges in designing efficient distributed rendezvous algorithms.

We have focused on the rendezvous problem between two users (Problem 5.1). We will then present algorithms for solving the $Alg - AS$ setting, in Chap. 6, where the users are allowed to run different algorithms. Then, we introduce rendezvous algorithms for the *Syn* setting in Chap. 7, where the users have to run symmetric algorithms and they start the rendezvous process at the same time. For the more general situations where the users can start the rendezvous process in different time slots, we first introduce algorithms for the *Anon* setting in Chap. 8 where the users have no distinguishable identifiers, and present rendezvous algorithms in Chap. 9 for the $Non - Anon$ setting, where the users have distinguishable labels. Finally, we show the method of extending rendezvous algorithms for two users to solving rendezvous among multiple users (Problem 5.2) in Chap. 10.

References

1. Kondareddy, Y., Agrawal, P., & Sivalingam, K. (2008). Cognitive radio network setup without a common control channel. In *MILCOM*.
2. Lazos, L., Liu, S., & Krunz, M. (2009). Spectrum opportunity-based control channel assignment in cognitive radio networks. In *SECON*.
3. Perez-Romero, J., Salient, O., Agusti, R., & Giupponi, L. (2007). A novel on-demand cognitive pilot channel enabling dynamic spectrum allocation. In *DySPAN*.
4. Stevenson, C. R., Chouinard, G., Lei, Z., Hu, W., Shellhammer, S. J., & Caldwell, W. (2009). IEEE 802.22: The first cognitive radio wireless regional area network standard. *IEEE Communications Magazine, 47*(1), 130–138.
5. Zhao, J., Zheng, H., & Yang, G.-H. (2005). Distributed coordination in dynamic spectrum allocation networks. In *DySPAN*.

Chapter 6
Asymmetric Blind Rendezvous Algorithms

Abstract In this chapter, we present asymmetric algorithms for the blind rendezvous problem. In the settings, we fix Alg as:

$$RS = < Alg - AS, Time, Port, ID, Non - Obli > \qquad (6.1)$$

where $Time \in \{Syn, Asyn\}$, $Port \in \{Port - S, Port - AS\}$, and $ID \in \{Non - Anon, Anon\}$. Although there are eight different rendezvous settings when Alg is fixed as asymmetric and $Label$ fixed as non-oblivious, in designing asymmetric algorithms, users' ID do not matter much. This is because the users' IDs are typically used to break symmetry in distributed computing, but we already assume the users can be distinguishable and they execute different algorithms. Therefore, we present how to design efficient algorithms for the four rendezvous settings (synchronous and port-symmetric, asynchronous and port-symmetric, synchronous and port-asymmetric, asynchronous and port-asymmetric), no matter the choice of ID from $\{Non - Anon, Anon\}$. In Sect. 6.1, we present two different types of rendezvous algorithms for the synchronous and port-symmetric rendezvous setting, and these algorithms are extended for the asynchronous and port-symmetric rendezvous setting in Sect. 6.2. In Sects. 6.3 and 6.4, we introduce efficient algorithms for the synchronous, port-asymmetric and asynchronous, port-asymmetric rendezvous settings. Finally, we summarize the chapter in Sect. 6.5.

6.1 Synchronous and Port-Symmetric Rendezvous

Consider two users u_i and u_j, suppose their available port sets are $C_i, C_j \subseteq U$ respectively. In the following settings:

$$RS = < Alg - AS, Syn, Port\text{-}S, ID, Non - Obli > \qquad (6.2)$$

two users have the same start time and both available port sets are the same: $C_i = C_j$.

© Springer Nature Singapore Pte Ltd. 2017
Z. Gu et al., *Rendezvous in Distributed Systems*,
DOI 10.1007/978-981-10-3680-4_6

6.1.1 Smallest Port Accessing Algorithm

Algorithm 6.1 Smallest Port Accessing Algorithm

1: Denote the set of available ports as $C \subseteq U$;
2: Find the smallest label $s \in C$ and access port s all the time;

This setting is the simplest one and two users can adopt the Smallest Port Accessing (SPA) algorithm (as shown in Algorithm 6.1) to achieve rendezvous. In the algorithm, the user chooses the port with the smallest label all the time. It is obvious that two users with symmetric port situations will rendezvous in their first attempt.

6.1.2 Quorum-Based Channel Hopping

Quorum-based Channel Hopping (QCH) [1, 2] generates the hopping sequence based on the quorum system which is defined in [1]:

Definition 6.1 Given a finite universal set $U = \{0, 1, \ldots, n - 1\}$ of n elements, a *quorum system S* under U is a collection of non-empty subsets of U, which satisfies the intersection property:

$$p \bigcap q \neq \emptyset, \forall p, q \in S \tag{6.3}$$

Each $p \in S$ (which is a subset of U) is called a quorum.

There are several ways of constructing a quorum system under set U and we will introduce a simple method called *cyclic quorum systems*, which is first introduced in [4]. To begin with, we introduce *relaxed cyclic difference set*.

Definition 6.2 A set $D = \{d_1, d_2, \ldots, d_k\} \subseteq U$ is called a relaxed cyclic (n, k)-difference set if for every $d \neq 0 \mod n$, there exists at least one ordered pair (d_i, d_j) where $d_i, d_j \in D$, such that $d_i - d_j \equiv d \pmod{n}$.

For example, if $n = 7, k = 3$, set $D = \{0, 1, 3\}$ is a relaxed cyclic $(7, 3)$-difference set under Z_7, where $Z_n = \{0, 1, \ldots, n-1\}$. Clearly, for any value $d \in \{1, 2, \ldots, 6\}$, there exist two elements in D that suit the equation. We define the cyclic quorum system as follows.

Definition 6.3 A group of sets $B_i = \{d_1 + i, d_2 + i, \ldots, d_k + i\} \mod n$, where $i \in \{0, 1, \ldots, n-1\}$ is a cyclic quorum system if and only if set $D = \{d_1, d_2, \ldots, d_k\}$ is a relaxed cyclic (n, k)-difference set.

We also use set $D = \{0, 1, 3\}$ as an example. We construct the cyclic quorum system as:

$$S = \{\{0, 1, 3\}, \{1, 2, 4\}, \{2, 3, 5\}, \{3, 4, 6\}, \{4, 5, 0\}, \{5, 6, 1\}, \{6, 0, 2\}\} \quad (6.4)$$

It is easy to check that any two elements in the quorum system S intersect. The QCH algorithm constructs different sequences on the basis of different quorums. Suppose there are N ports $\{1, 2, \ldots, N\}$ and there exists a cyclic quorum system $S = \{B_0, B_1, \ldots, B_{n-1}\}$ under Z_n. The QCH algorithm constructs sequence S_i for each quorum B_i as follows.

(1) Step 1: Denote $B_i = \{d_1, d_2, \ldots, d_k\}$;
(2) Step 2: For each port $1 \le j \le N$, construct a frame of N time slots $\{u_0, u_1, \ldots, u_{n-1}\}$ as:

$$u_l = \begin{cases} j & \text{if } l \in B_i \\ * & \text{otherwise} \end{cases} \quad (6.5)$$

where $*$ can be any port in set $\{1, 2, \ldots, N\}$.
(3) Step 3: The constructed sequence S_i is composed of such N frames and each frame consists of n elements.

For example, there are three ports $\{1, 2, 3\}$ and we can construct a cyclic quorum system under Z_3 as:

$$S = \{\{0, 1\}, \{1, 2\}, \{2, 0\}\} \quad (6.6)$$

For each quorum in set S, we construct the corresponding sequences. For quorum $\{0, 1\}$, we construct the sequence as:

$$S_1 = \{11 * |22 * |33*\} \quad (6.7)$$

where the symbol | separates different frames and $*$ is any port in $\{1, 2, \ldots, N\}$. Similarly, we can construct the other two sequences as:

$$S_2 = \{*11| * 22| * 33\}$$
$$S_3 = \{1 * 1|2 * 2|3 * 3\} \quad (6.8)$$

For two different users u_a and u_b, they can choose different quorums in the constructed cyclic quorum system and they would choose the ports for rendezvous attempt according to the corresponding sequence. For example, user u_a chooses the constructed sequence S_1 to access the port periodically while user u_b chooses sequence S_2 for rendezvous. According to the definition of cyclic quorum system, it is easy to see that the corresponding quorums should intersect and the corresponding choice in the sequence should be the same port, which implies rendezvous. Therefore, the QCH algorithm can guarantee rendezvous for two synchronous users.

Notice that, the QCH algorithm is designed for the special situation that all port are available. By a small modification, it can be applied to the scenario that two users have symmetric available ports.

Suppose the available ports for the symmetric users are $C = \{p_1, p_2, \ldots, p_n\} \subseteq U$, which implies there are n available ports for the users. We reconstruct set $C' = \{1, 2, \ldots, n\}$ to apply the QCH algorithm. When we need to choose port i in set C', we replace it with port p_i in set C, which can be used to guarantee rendezvous in the port symmetric situation.

6.2 Asynchronous and Port-Symmetric Rendezvous

Consider two users u_i and u_j, suppose their available port sets are $C_i, C_j \subseteq U$. In the following settings:

$$RS = < Alg - AS, Asyn, Port\text{-}S, ID, Non - Obli > \tag{6.9}$$

two users start the rendezvous process in different time slots and both sets are the same, i.e. $C_i = C_j$.

6.2.1 Asynchronous Quorum-Based Channel Hopping

Asynchronous QCH (A-QCH) [3] is modified for asynchronous users, but only applicable to two available channels. We describe the A-QCH algorithm briefly and readers may refer to [3] for more details.

The QCH algorithm in Sect. 6.1.2 cannot be applied to two asynchronous users, because two users choosing different quorums p, q in a quorum system have clock skew; we can consider the situation as one user is adopting the rotated quorum by some bias, such as $rotate(q, k)$, which means each element in quorum q rotates k numbers. Then, quorum p and $rotate(q, k)$ may not intersect. The modification in A-QCH uses two cyclic quorum systems to construct such port accessing sequence, but it only works for two available ports.

Denote two available port as $P = \{p_0, p_1\}$, and suppose there are n time slots in each constructed frame. The algorithm works as follows:

(1) Denote the set Z_n as $\{0, 1, \ldots, n - 1\}$;
(2) Find a minimal (n, k) cyclic difference set $D = \{d_1, d_2, \ldots, d_k\}$ under Z_n such that $k < \frac{n}{2}$;
(3) Construct the minimal cyclic quorum system $S = \{B_i | B_i = \{d_1 + i, d_2 + i, \ldots, d_k + i\} \mod n$ where $i \in [0, n - 1]$;
(4) Find a relaxed (n, k') cyclic different set $D' = \{d'_1, d'_2, \ldots d'_{k'}\}$ under Z_n where $k' = \lceil \frac{n+1}{2} \rceil$) and $D' \bigcap D = \emptyset$;

Fig. 6.1 An example of the
A-QCH algorithm

p_0	p_0	p_0	p_1	p_0	p_1	p_1	p_1	p_1
p_1	p_0	p_0	p_0	p_1	p_0	p_1	p_1	p_1
p_1	p_1	p_0	p_0	p_0	p_1	p_0	p_1	p_1
p_1	p_1	p_1	p_0	p_0	p_0	p_1	p_0	p_1
p_1	p_1	p_1	p_1	p_0	p_0	p_0	p_1	p_0
p_0	p_1	p_1	p_1	p_1	p_0	p_0	p_0	p_1
p_1	p_0	p_1	p_1	p_1	p_1	p_0	p_0	p_0
p_0	p_1	p_0	p_1	p_1	p_1	p_1	p_0	p_0
p_0	p_0	p_1	p_0	p_1	p_1	p_1	p_1	p_0

(5) Construct the cyclic quorum system $S' = \{B_i'|B_i' = \{d_1' + i, d_2' + i, \ldots, d_{k'}' + i\}$ mod n where $i \in [0, n-1]$;

(6) Construct the sequence with n frames and each frame contains n elements;

(7) For the jth frame, the ith element, we assign the port as:

$$s_{ji} = \begin{cases} p_0 & if\ i \in B_j \\ p_1 & if\ i \in B_j' \\ * & otherwise \end{cases} \qquad (6.10)$$

where $*$ can be any port.

(8) The user accesses the port according to the constructed sequence periodically.

The method of constructing minimal (n, k) cyclic difference set and relaxed (n, k') cyclic different set can be found in [12] and we do not introduce the details. For example, if $n = 9$, and we construct set $D = \{0, 1, 2, 4\}$ and set $D' = \{3, 5, 6, 7, 8\}$. It is easy to check that both sets are relaxed cyclic difference set and $D \cap D' = \emptyset$. Then, we can construct the sequence as in Fig. 6.1, where there are 9 frames and each frame contains 9 elements.

Two different users compute different relaxed difference sets and the constructed sequences are different. However, by involving two quorum systems, two different users can always achieve rendezvous on the port p_0 or p_1 (notice that two symmetric users should have at least two available ports p_0, p_1 to execute the algorithm).

6.2.2 Sequential Accessing Algorithm

We propose the Sequential Accessing Algorithm in Algorithm 6.2. In the algorithm, we first count the number of elements in the available port set as n. In each time slot t, we compute the xth element in set C where x is t's modulus under n. This is similar to accessing the available ports sequentially from the 1th label to the nth label. When t is larger than n, we repeat the sequential accessing.

Algorithm 6.2 Sequential Accessing Algorithm

1: Denote time $t := 1$, the user's port set as $C \subseteq U$;
2: Denote the cardinality as $n := |C|$;
3: **while** Not rendezvous **do**
4:     Let $x := (t - 1)\%n + 1$;
5:     Let p_{id} be the xth number in set C;
6:     Access port p_{id} in time t;
7:     $t := t + 1$;
8: **end while**

If two users are port-symmetric, but asynchronous, suppose one user u_i runs Algorithm 6.2 while user u_j runs a simple algorithm modified from Algorithm 6.1: user u_j chooses a label in set C_j and access the port all the time. It is easy to show that two users can rendezvous within $O(n)$ time slots.

Theorem 6.1 *Two port-symmetric, asynchronous users running Algorithm 6.2 and modified Algorithm 6.1 can rendezvous in $O(|C|)$ time slots, where C is the set of available ports.*

Proof Suppose user u_i starts Algorithm 6.2 later than user u_j. Suppose user u_j chooses the kth label in its available port set C_j, where $1 \leq k \leq |C_j|$. Obviously, when user u_i starts the algorithm, it can achieve rendezvous in k time slots, from user u_i's clock.

Supposing user u_i starts earlier than user u_j, when user u_j starts accessing the kth port, it may not achieve rendezvous with user u_i quickly. However, since user u_i repeats accessing the ports sequentially, it will definitely access the kth port within $|C_i|$ time slots.

Combining these two aspects, the theorem holds.

As illustrated in Fig. 6.2, the available port sets for two users are $\{1, 2, 7\}$; user u_i runs Algorithm 6.2 while user u_j runs the modified Algorithm 6.1 by accessing port 7. As shown in the figure, when user u_i starts earlier (as Fig. 6.2a) or later (as Fig. 6.2b) than user u_j, they can all achieve rendezvous in 3 time slots.

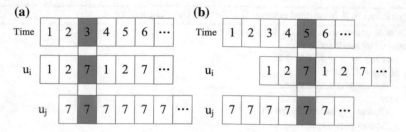

Fig. 6.2 Rendezvous examples when user u_i runs Algorithm 6.2 while user u_j runs the modified Algorithm 6.1

Actually, if both users are aware of the port-symmetric situation, they can also run a symmetric algorithm for rendezvous. Suppose both users adopt Algorithm 6.1 designed for synchronous and port-symmetric rendezvous. Without loss of generality, suppose user u_i starts $\Delta > 0$ time slots earlier than user u_j. When user u_j starts the rendezvous process at time $\Delta + 1$ (from user u_i's clock), it will access port s (the smallest port) in the first time slot (from user u_j's clock). As user u_i will always access port s, they could rendezvous in their first rendezvous attempt. Thus, the time to rendezvous is 1, where TTR is defined as the cost time to rendezvous for the user who starts latter in Problem 5.1.

Although port-symmetry is a easy situation to handle, the users are not aware of the situation and whether they are symmetric or not. Therefore, Algorithm 6.1 cannot work if two users have asymmetric ports. Therefore, we hope to design efficient algorithms that work for the asymmetric port situation, while it also has good performance when the ports are symmetric. We will introduce such algorithms in the following sections.

6.3 Synchronous and Port-Asymmetric Rendezvous

Consider two users u_i and u_j, and suppose their available port sets are $C_i, C_j \subseteq U$. In the following settings:

$$RS = < Alg - AS, Syn, Port\text{-}AS, ID, Non - Obli > \qquad (6.11)$$

where two users start the rendezvous process at the same time but the sets of available ports may be different, i.e. $C_i \neq C_j$.

Algorithm 6.3 Modified Sequential Accessing Algorithm

1: Denote time $t := 1$, the user's port set as $C \subseteq U$;
2: **while** Not rendezvous **do**
3:    Let $p_{id} := (t-1)\%N + 1$;
4:    **if** $p_{id} \in C$ **then**
5:       Access port p_{id} in time t;
6:    **else**
7:       Choose p_{id} randomly from C;
8:       Access port p_{id} in time t;
9:    **end if**
10:   $t := t + 1$;
11: **end while**

6.3.1 Modified Sequential Accessing Algorithm

We present the Modified Sequential Accessing (MSA) Algorithm as described as Algorithm 6.3. First of all, the user computes the port with id p_{id} corresponding to the current time slot t as $p_{id} = (t-1)\%N + 1$. Clearly, it is similar to accessing the ports by repeating the sequence $\{1, 2, \ldots, N\}$. However, due to occupancy by other services, some ports are not available for the user. Thus, it needs to select another available port randomly from set C. We show that, users u_i and u_j running Algorithm 6.2 can always achieve rendezvous within N time slots.

Theorem 6.2 *The synchronous users u_i and u_j can achieve rendezvous within N time slots, by running Algorithm 6.2 at the same time.*

Proof For the two neighboring users u_i and u_j, their sets of available ports must intersect to ensure at least one common available port exists. Therefore $C_i \cap C_j \neq \emptyset$. Denote the smallest number in set $C_i \cap C_j$ as s, clearly, $1 \leq s \leq N$.

When two users run the algorithm at the same time, when $t = s$, port s is available for user u_i since $s \in C_i$, and thus user u_i should access port s. Similarly, user u_j will access port s since it is available. Therefore, two users can access the connected ports and they rendezvous in time slot s. So the theorem holds.

6.4 Asynchronous and Port-Asymmetric Rendezvous

Consider two users u_i and u_j, and suppose their available port sets are $C_i, C_j \subseteq U$. In the following settings:

$$RS = <Alg - AS, ASyn, Port\text{-}AS, ID, Non - Obli> \qquad (6.12)$$

where two users start the rendezvous process in different time slots and the sets of available ports may be different, i.e. $C_i \neq C_j$. This situation is the most difficult one in this chapter and we present some elegant results.

6.4.1 Sequential Access and Temporary Wait for Rendezvous

We present the Temporary Wait algorithm as in Algorithm 6.4. This algorithm works in this fashion: for each time slot t, compute the corresponding value x within range $[1, 2N^2]$ as $x := (t-1)\%2N^2+1$. We can think of this operation as repeating the time every $2N^2$ time slots. Following that, we divide the $2N^2$ time slots into N frames and each frame contains $2N$ time slots. This is why we compute $p_{id} := \lceil(x-1)/(2N)\rceil+1$ (p_{id} corresponds to the frame that time slot t belongs to). Similar to the Modified Sequential Accessing Algorithm, if port p_{id} is not available, we will choose a random available port as a replacement. This process continues until rendezvous.

Algorithm 6.4 Temporary Wait Algorithm

1: Denote time $t := 1$, the user's port set as $C \subseteq U$;
2: **while** Not rendezvous **do**
3:    Let $x := (t - 1)\%2N^2 + 1$;
4:    Let $p_{id} := \lceil(x - 1)/(2N)\rceil + 1$;
5:    **if** p_{id} does not belong to set C **then**
6:       Choose p_{id} as a random value from C;
7:    **end if**
8:    Access port p_{id} for rendezvous attempt;
9:    $t := t + 1$;
10: **end while**

We present a clear illustration in Fig. 6.3. The algorithm will access a fixed port for $2N$ time slots (if we do not consider the situation that it is not available and should be replaced). Then, after every $2N$ time slots, the algorithm will choose the next port for waiting (also over $2N$ time slots). And this is the reason we call it the Temporary Wait Algorithm. As shown in the figure, the user accesses a fixed port for $2N$ time slots, and the replacement happens if some port is not available. For example, port k replaces port 2 in the figure, if port 2 is not in the user's available port set.

For two users u_i and u_j, suppose one user (without loss of generality, u_i) adopts the Modified Sequential Accessing Algorithm (Algorithm 6.3) while the other user u_j runs the Temporary Wait Algorithm (Algorithm 6.4). We show that they can achieve rendezvous within $2N^2$ time slots.

Theorem 6.3 *Two users, adopting Algorithms 6.3 and 6.4 respectively, can achieve rendezvous within $MTTR = 2N^2 = O(N^2)$ time slots.*

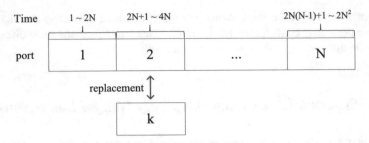

Fig. 6.3 The illustration of the Temporary Wait Algorithm

Proof For two neighboring users u_i and u_j, their sets of available ports must intersect to ensure at least one common available port exists. Therefore $C_i \cap C_j \neq \emptyset$. Denote the smallest number in set $C_i \cap C_j$ as s, clearly, $1 \le s \le N$.

First, we show that, supposing user u_j waits on port s for $2N$ time slots from $t+1$ to $t+2N$, if user u_i has begun Algorithm 6.3 no later than $t+N$, they can always achieve rendezvous. Actually, if user u_i starts the algorithm at time $t_i \le t+N$, during the time slots from t_i to $t_i + N - 1$, user u_j will wait on port s, while user u_i will access port $\{1, 2, \ldots, s, \ldots, N\}$ sequentially (notice that user u_i does not access the unavailable ports, but it does affect the analysis since s is available). Therefore, they must rendezvous within these N time slots.

Then, we analyze the impact of asynchronous start. If user u_i starts the algorithm earlier than user u_j, it is clear that they can achieve rendezvous when user u_j waits on choosing port s, thus $TTR \le s * 2N \le 2N^2$. If user u_i starts later, the worst situation would happen when user u_j is finishing waiting on port s but u_i just starts. Considering any $2N$ time slots that user u_j waits on port s, denote them as $t+1$ to $t+2N$. If user u_i starts at time $t+2N-s+1$, user u_i will choose port s at time $t+2N+1$ but user u_j has just moved to the next port for waiting. However, after $2N * (N-1)$ time slots, i.e. from $t+N^2+1$ to $t+N^2+2N$, user u_j will also access port s and they will rendezvous in the first N time slots. Then, we can conclude that time to rendezvous is also bounded by $2N^2$ time slots if user u_i starts later.

Combining these, two users running asymmetric algorithms can achieve rendezvous within $2N^2 = O(N^2)$ time slots.

6.5 Chapter Summary

In this chapter, we present different types of rendezvous algorithms when the users can run asymmetric algorithms. In practical applications, the users in the distributed system may have different roles in the communications. For example, if one node in the system tries to broadcast a message to all neighboring nodes, it may play the role of "sender", while the other nodes who do not send messages are regarded as "receiver". For example, wireless sensor network (WSN) is a typical distributed

Table 6.1 Rendezvous algorithms for different rendezvous settings

Algorithms	Synchronous	Asynchronous
Port-symmetric	SPA, QCH	A-QCH, SAA
Port-asymmetric	MSA	TWA

system where the sensors can have different roles in constructing communication links. Normally, each sensor node can send or receive signals through radio (bidirectional or unidirectional radios). Suppose all sensor nodes are deployed in a monitoring area where they can sense environmental data. Further suppose there exists a mobile sink (it can be a moving vehicle which carries sensors or communication units) that travels through the monitoring area; when it is close to some deployed sensor nodes, it can send signals to activate these sensors and then collect data from them. In this case, the mobile sink can be regarded as the "sender", while the deployed sensors are "receivers". Therefore, they can execute different algorithms to establish communication links.

In this chapter, we mainly introduce algorithms for four different rendezvous settings: *synchronous and port-symmetric*, *asynchronous and port-symmetric*, *synchronous and port-asymmetric*, and *asynchronous and port-asymmetric*. Since the users' IDs are used to break symmetry among the users, we do not consider the impact of users' IDs in the chapter.

For the synchronous and port-symmetric setting, we present two rendezvous algorithms that can perform well. The first one (Smallest Port Accessing algorithm, SPA) simply accesses the port with smallest label, while the second one (quorum-based channel hopping, QCH) adopts a quorum system to design efficient port accessing strategy. The SPA algorithm can be used in very limited situations, but the QCH algorithm can be applied in many rendezvous settings.

For the asynchronous and port-asymmetric setting, we present the method of extending the QCH algorithm to two asynchronous users. The asynchronous QCH (A-QCH) algorithm designs special rendezvous sequences on the basis of two disjoint quorum systems and this method can also be applied in designing symmetric algorithms for the users. Another algorithm is called the Sequential Accessing Algorithm (SAA), which accesses the ports in a sequential manner. But this algorithm has limited extensions.

For the synchronous and port-asymmetric setting, we propose the Modified Sequential Accessing (MSA) algorithm which operates on the basis of the SAA algorithm. When one user adopts the MSA algorithm while the other one runs the SPA algorithm, they can rendezvous in a short time.

For the asynchronous and port-asymmetric setting, one user adopts the MSA algorithm while the other user adopts the Temporary Wait Algorithm (TWA), and they can achieve rendezvous quickly. The intuitive idea is one user keeps accessing the ports dynamically, while the other one moves slowly enough such that the first user can peruse all the ports during the "slow" moves of the other user.

From these rendezvous algorithms, the main idea in designing asymmetric algorithms is to make one user wait on a fixed port for a sufficient amount of time, while the other user keep accessing the ports dynamically. The described algorithms are listed in Table 6.1 and readers who are interested in asymmetric algorithms can design some other algorithms for rendezvous.

References

1. Bian, K., Park, J.-M., & Chen, R. (2009). A quorum-based framework for establishing control channels in dynamic spectrum access networks. In *Mobicom*.
2. Bian, K., Park, J.-M., & Chen, R. (2011). Control channel establishment in cognitive radio networks using channel hopping. *IEEE Journal on Selected Areas in Communications*, *29*(4), 689–703.
3. Bian, K., & Park, J.-M. (2013). Maximizing Rendezvous diversity in rendezvous protocols for decentralized cognitive radio networks. *IEEE Transactions on Mobile Computing*, *12*(7), 1294–1307.
4. Luk, W. S., & Wong, T. T. (1997). Two new quorum based algorithms for distributed mutual exclusion. In *ICDCS*.

Chapter 7
Synchronous Blind Rendezvous Algorithms

Abstract In this chapter, we present *symmetric algorithms* for the blind rendezvous problem of two synchronous users. In the settings, we fix Alg and $Time$ as:

$$RS =< Alg\text{-}S, Syn, Port, ID, Non\text{-}Obli > \qquad (7.1)$$

where $Port \in \{Port - S, Port - AS\}$ and $ID \in \{Non - Anon, Anon\}$. As shown in the previous chapter, designing asymmetric rendezvous algorithms for synchronous users is relatively simple. In this chapter, we show that designing symmetric rendezvous algorithms for synchronous users is also not that difficult, regardless of whether they are port-symmetric or port-asymmetric, and anonymous or non-anonymous. In Sect. 7.2, we present a rendezvous algorithm called the expanded sequential accessing algorithm, which can work efficiently for synchronous users who run symmetric algorithms. Although this algorithm can only work for synchronous users, the intuitive idea behind can be used in many other situations. We summarize the chapter in Sect. 7.2.

7.1 Expanded Sequential Accessing Algorithm

It is easy to see that there are four different rendezvous settings when Alg is fixed as symmetric, $Time$ is fixed as synchronous, and $Label$ is fixed as non-oblivious. Recall the proposed algorithms in Chap. 6 that although we are designing asymmetric algorithms, it is much easier for two synchronous users to rendezvous. In addition, the time complexity of such rendezvous algorithms for synchronous users is low ($O(N)$ time slots).

In this section, we introduce some special constructions for synchronous users and we omit the impact of users' IDs. The Expanded Sequential Accessing (ESA) Algorithm is described in Algorithm 7.1. We design the algorithm on the basis of the Modified Sequential Accessing Algorithm presented in the previous chapter, where each user accesses the ports in a sequential order from 1 to N. Different from that, we repeat the rendezvous process every $2N$ time slots. We can consider the operation as expanding each time slot of the Sequential Accessing algorithm into

© Springer Nature Singapore Pte Ltd. 2017
Z. Gu et al., *Rendezvous in Distributed Systems*,
DOI 10.1007/978-981-10-3680-4_7

Algorithm 7.1 Expanded Sequential Accessing Algorithm

1: Denote time $t := 1$, the user's port set as $C \subseteq U$;
2: Choose the smallest label is C as s;
3: **while** Not rendezvous **do**
4: **if** $t\%2 == 1$ **then**
5:    Let $f_{id} := s$;
6: **else**
7:    Let $p_{id} := (t/2 - 1)\%N + 1$;
8:    **if** p_{id} does not belong to set C **then**
9:      Let p_{id} be a random number from set C;
10:    **end if**
11: **end if**
12: Access port p_{id} in time t;
13: $t := t + 1$;
14: **end while**

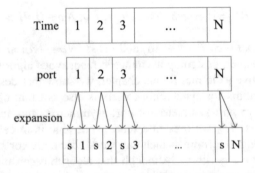

Fig. 7.1 The illustration of the Expanded Sequential Accessing Algorithm

two slots, where the first slot is for waiting on a fixed port s, the smallest port, and the second slot is for keeping to the same rule as before, accessing the ports sequentially, $\{1, 2, \ldots, N\}$. We illustrate the construction in Fig. 7.1. Similar to the introduced rendezvous algorithms for asymmetric users, if some port is not available, we replace it by a random available one.

The advantage of the Expanded Sequential Accessing algorithm is that it guarantees fast rendezvous between two synchronous users, no matter whether the scenario is port-symmetric or port-asymmetric. We do not consider the users' IDs here since the time complexity is already low. We derive the time complexity of the Modified Sequential Accessing algorithm for both the $Port - S$ and $Port - AS$ settings.

Theorem 7.1 *Two users running the Expanded Sequential Accessing algorithm (Algorithm 7.1) can achieve rendezvous in the first time slot if they are port-symmetric, i.e.* $RS = <$ Alg-S, Syn, Port-S, ID, Non-Obli $>$, *where ID can be Anon or* $Non - Anon$.

Proof Considering two users u_i and u_j, denote their available port sets as C_i, C_j respectively. In the synchronous setting, both users start the rendezvous algorithm at the same time. When two users are port-symmetric, i.e. $C_i = C_j$, we show that they can rendezvous very quickly.

According to Algorithm 7.1, user u_i will choose the smallest port in set C_i (denote it as s_i) in the first time slot, while user u_j will also choose the smallest port in set C_j (denote it as s_j). Since $C_i = C_j$, $s_i = s_j$, they can rendezvous in the first time slot.

Theorem 7.2 *Two users running the Expanded Sequential Accessing algorithm (Algorithm 7.1) can achieve rendezvous in $TTR \leq 2N = O(N)$ time slots if they are port-asymmetric, i.e. $RS =< $ Alg-S, Syn, Port-AS, ID, Non-Obli $>$, where ID can be Anon or $Non - Anon$.*

Proof Considering two users u_i and u_j, denote their available port sets as C_i, C_j respectively. In the synchronous setting, both users start the rendezvous algorithm at the same time. When two users are port-asymmetric, i.e. $C_i \neq C_j$, we show that they can rendezvous in $O(N)$ time slots.

Denote the smallest port of C_i as s_i and the smallest port of C_j as s_j. Since $C_i \neq C_j$, these two ports may be different, i.e. $s_i \neq s_j$. Therefore, when the time slot is an odd number, two users may not achieve rendezvous. However, when the time slot t is an even number in $[2, 2N]$, they can rendezvous. Denote $G = C_i \bigcap C_j$ and the smallest port in G as s_g. Clearly, in time slot $2s_g$, both users u_i and u_j will choose port s_g according to the algorithm. Thus, they can rendezvous within $TTR = 2s_g \leq 2N = O(N)$ time slots.

Remark 7.1 The method of dividing a time slot into multiple time slots can help design efficient rendezvous algorithms. In this chapter, we only show the time division method for rendezvous between two synchronous users. We will show some other constructions, such as rendezvous for asynchronous users.

7.2 Chapter Summary

In this chapter, we start to handle the blind rendezvous problem if the users have to run symmetric algorithms. Different from the previous chapter, all nodes in the distributed system do not have special roles and they have to run a symmetric algorithm.

In order to understand the idea of designing symmetric rendezvous algorithms, we study the simplest setting where the users start the rendezvous process at the same time. No matter whether the users have symmetric or asymmetric available ports, we introduce a time division method where one slot is designed for regular rendezvous attempt, while the other slot can handle some special rendezvous construction. In our designed Expanded Sequential Accessing algorithm, if the users have symmetric ports, they can achieve rendezvous in the first time slot; if the users have asymmetric ports, they can also rendezvous in $O(N)$ time slots. This time division method serves as a good technique for rendezvous.

Notice that, in the previous chapter, we design different rendezvous algorithms for different rendezvous settings, and one particular algorithm cannot be applied to the other settings. However, in this chapter, our designed algorithm is applicable to all the four rendezvous settings, and this is a major improvement over the previous chapter. In the following chapters, we will show how to design such unified rendezvous algorithms that can work for many rendezvous settings.

Chapter 8
Asynchronous Blind Rendezvous Algorithms for Anonymous Users

Abstract In this chapter, we present *symmetric algorithms* for the blind rendezvous problem between two *asynchronous, anonymous* users. In the rendezvous setting, we fix Alg, $Time$, and ID as follows:

$$RS = < Alg\text{-}S, Asyn, Port, Anon, Non\text{-}Obli > \tag{8.1}$$

where $Port \in \{Port - S, Port - AS\}$, which implies that we will design efficient algorithms that have good performance for both symmetric and asymmetric port situations. In this chapter, we will introduce a commonly used technique in designing rendezvous algorithms for cognitive radio networks, which is called Channel Hopping (CH) [1, 2, 11, 13, 14]. The intuitive idea is: in order to guarantee rendezvous for asynchronous users, the rule to access the licensed channels (in the network) should be periodic. Thus, we should construct a sequence of fixed length, such as $S = \{s_0, s_1, \ldots, s_{T-1}\}$ where s_i is an available channel and the user hops among the channels by repeating the sequence, i.e. they access $s_{t \bmod T}$ at time t. Rendezvous in the distributed system is similar to rendezvous in the cognitive radio networks, and we can use the Channel Hopping technique to design efficient algorithms. In a distributed system, the available port sets for asymmetric users could be different, and different users may construct different hopping sequences. Therefore, it is difficult to design efficient algorithms (or short hopping sequences) that are suitable for all users. Moreover, the lower bound of such sequence cannot be derived directly when the sequences for different users vary, which is important for evaluating and verifying the efficiency of any proposed rendezvous algorithm. Therefore, we introduce *Global Sequence (GS) based rendezvous algorithms* to alleviate the impact of asymmetry in the ports' occupancy; the intuitive idea is: design a fixed sequence $S = \{s_0, s_1, \ldots, s_{T-1}\}$ for all users based on the full port set $U = \{1, 2, \ldots, N\}$ and each user hops among the ports by repeating the sequence (modification on the sequence may exist when some ports are not available for communication). Specifically, the GS based rendezvous algorithms work in two phases:

Phase 1: Assume all users have the same available port set U and design the GS on the basis of U;

© Springer Nature Singapore Pte Ltd. 2017

Z. Gu et al., *Rendezvous in Distributed Systems*,

DOI 10.1007/978-981-10-3680-4_8

Phase 2: Each user modifies the sequence according to its own available port set, i.e. when the user should access an unavailable port by the original hopping sequence, replace it with an available one that is picked randomly or by some pre-defined rules.

In this chapter, we introduce efficient GS based algorithms which can guarantee rendezvous for both symmetric and asymmetric users in a short time. To begin with, we introduce two simple algorithms for two asynchronous users that are port-symmetric in Sects. 8.1 and 8.2. Then, we introduce three GS based algorithms that have good performance for both port-symmetric and port-asymmetric scenarios. We introduce the Channel Rendezvous Sequence (CRSEQ) algorithm in Sect. 8.3, the Jump Stay (JS) algorithm in Sect. 8.4, and the Disjoint Relax Different Set (DRDS) based algorithm in Sect. 8.5. We also show the lower bound of such a GS based rendezvous algorithm in Sect. 8.6. We summarize the chapter in Sect. 8.7.

8.1   Generated Orthogonal Sequence (GOS)

Generated Orthogonal Sequence (GOS) [5] is considered pioneering work in cognitive radio networks, which generates a hopping sequence of length $N(N + 1)$ on the basis of a random permutation of the set $\{1, 2, \ldots, N\}$. Technically, a random permutation of $\{1, 2, \ldots, N\}$ is chosen from the $N!$ permutations. Then the GOS is constructed as follows:

(1) Denote the random permutation of $\{1, 2, \ldots, N\}$ as $\{k_1, k_2, \ldots, k_N\}$;
(2) the GOS consists of N phases where each phase contains $N + 1$ elements;
(3) for phase i, construct the phase as $\{k_i, k_1, k_2, \ldots, k_N\}$;

We can regard this process as embedding the permutation N times within a super-sequence of the permutation. Figure 8.1 depicts the example of the construction, where a permutation $\{2, 4, 1, 3\}$ is selected when $N = 4$, and the GOS sequence is

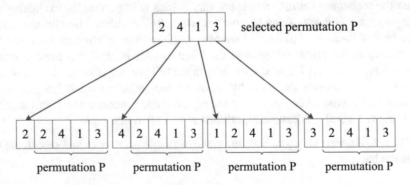

Fig. 8.1   An example of the Generated Orthogonal Sequence

constructed by the steps introduced. However, this algorithm is limited to the situation that all channels are available. We show the correctness of GOS briefly.

First, if two users repeat the same GOS at the same time, rendezvous happens in the first time slot. So we only need to consider two asynchronous users hopping with the same GOS. Without loss of generality, assume user u_i is δ time slots earlier than user u_j, and then there are two situations according to different δ values:

(1) $\delta \% (N + 1) = 0$. It is easy to check that in the first phase of user u_j, it runs the same sequence of length N except the first number because they all use the same permutation $\{k_1, k_2, \ldots, k_N\}$ after k_i for the i-th phase;

(2) $\delta \% (N + 1) \neq 0$. The first number of the N phases of user u_j will meet every number in the permutation $\{k_1, k_2, \ldots, k_N\}$ and thus rendezvous is guaranteed.

Combining these two situations, the GOS can guarantee rendezvous between two users if all channels are available.

8.2 Deterministic Rendezvous Sequence (DRSEQ)

GOS can guarantee rendezvous between two asynchronous users under the situation that the all ports are available (it is easy to extend the algorithm to a distributed system), which is a very special port-symmetric situation. Following this work, *Deterministic Rendezvous Sequence (DRSEQ)* of length $2N + 1$ is proposed in [15], which works better under the situation that all ports are available. The main idea of the algorithm is to construct a simple sequence as:

$$\{1, 2, \ldots, N, e, N, N - 1, \ldots, 1\} \tag{8.2}$$

where e means the user can choose no (or any) port at this moment. Figure 8.2 shows an example of the rendezvous situations when $N = 4$. Supposing user u_i starts at time 0, the constructed sequence is:

$$\{1, 2, 3, 4, e, 4, 3, 2, 1\} \tag{8.3}$$

When user u_j starts the algorithm, we suppose it is δ time slots later than user u_i. When $\delta \in [0, 8]$, we list the rendezvous situations in the figure.

The correctness can be verified. Denote the constructed sequence as $\{a_0, a_1, \ldots, a_{2N}\}$ and each element in the sequence is constructed by the following equations:

$$a_i = \begin{cases} i + 1 & \text{when } 0 \leq i < N \\ e & \text{when } i = N \\ 2N + 1 - i & \text{when } N + 1 \leq i \leq 2N + 1 \end{cases} \tag{8.4}$$

Therefore, when one user is δ time slots earlier than the other, we can use the equations to compute the rendezvous port easily. For example, when $\delta = 1$, suppose

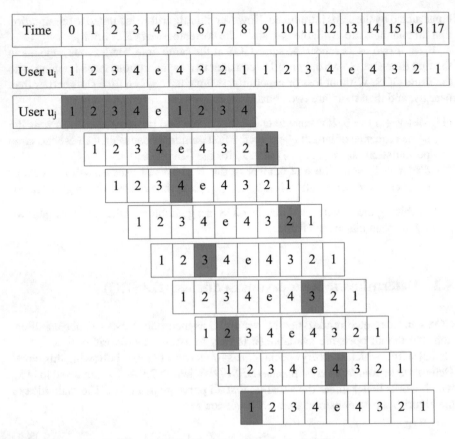

Fig. 8.2 An example of Deterministic Rendezvous Sequence when $N = 4$

rendezvous happens at time t. If $0 \leq t \leq N$, we can check that $a_t \neq a_{t+1}$. If $N + 1 \leq i \leq 2N$, $a_t \neq a_{t+1}$ since $N + 1 \leq t + 1 \leq 2N + 1$. Then when $t = 2N + 1$, $a_t = 1$ and $a_{t+1} = a_1 = 1$. Therefore, rendezvous happens on port 1 and the time to rendezvous is $2N + 1$.

The algorithm can guarantee rendezvous between two asynchronous users no matter what value δ is equal to. The readers could derive the time to rendezvous of various δ values.

8.3   Channel Rendezvous Sequence (CRSEQ) Algorithm

Channel Rendezvous Sequence (CRSEQ) [13] is the first algorithm guaranteeing rendezvous in a bounded time when only a portion of the channels are available in a cognitive radio network. The main technique is to design a global sequence based on the triangle number. There are P periods in constructing the CRSEQ, where P is the smallest prime number suiting $P > N$ and each period consists of

Table 8.1 MTTR Comparison for GS based rendezvous algorithms

Algorithms	Symmetric	Asymmetric
GOS [5]	$N(N+1) = O(N^2)$	–
DRSEQ [15]	$2N + 1 = O(N)$	–
CRSEQ [13]	$P(3P - 1) = O(N^2)$	$P(3P - 1) = O(N^2)$
Jump-Stay [11]	$3P = O(N)$	$3NP(P - G) = O(N^3)$
EJS [10]	$O(N)$	$O(N^2)$
DRDS [6]	$3P = O(N)$	$3P^2 + 2P = O(N^2)$

Remarks: (1) "–" means DRSEQ and GOS are inapplicable to asymmetric users; (2) P is the smallest prime number no less than N, $P = O(N)$

$3P - 1$ number. For the i-th period where $i \in [1, P]$, denote the triangle number as $T_i = \frac{i(i+1)}{2}$ and the constructed period as $\{a_{i,0}, a_{i,1}, \ldots, a_{i,3P-2}\}$, the period can be constructed according to the following equations:

$$a_{i,j} = \begin{cases} T_i + j \mod P + 1 & \text{when } 0 \leq j < 2P - 1 \\ \lfloor \frac{i}{3P-1} \rfloor \mod P + 1 & \text{when } N + 1 \leq i \leq 2N + 1 \end{cases} \tag{8.5}$$

CRSEQ can guarantee rendezvous for two users in $3P^2$ time slots when the users share some common available channels. However, it works badly when the users are symmetric as shown in Table 8.1. We omit the proof of the correctness and the reader may refer to [13] for more details.

8.4 Jump Stay Algorithm

Jump-Stay (JS) [11] is another efficient algorithm which guarantees fast rendezvous between symmetric users in cognitive radio networks. The main idea is similar to CRSEQ, which generates the global sequence of P periods and each period contains two *jump* frames and one *stay* frame (each frame contains P numbers, where P is the smallest prime number larger than the number of all licensed channels N).

We describe the two types of frames as follows. Denote the starting index as i and the step length as r. In the jump frame, the j-th number (denote as a_j) is computed by:

$$a_j = (i + r * j - 1) \mod P + 1 \tag{8.6}$$

and each frame contains P numbers. In the stay frame, the users stays at channel i for P time slots.

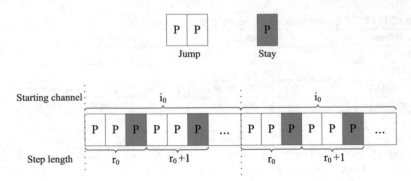

Fig. 8.3 Jump Stay Algorithm

We describe the JS algorithm as follows:

(1) Compute P as the smallest prime number larger than the number of all channels N;
(2) denote the initiate starting channel as $i_0 \in [1, N]$ and initiate step length $r_0 \in [1, p)$;
(3) the sequence is composed of N rounds, and the starting index is $i_0 + k$ in the k-th round;
(4) in the first round, the starting index keeps at channel i_0 and there are N loops inside each round;
(5) in the i-th loop, the step length is $r_0 + i$ and three frames are constructed as *Jump, Jump, Stay* on the basis of starting index and step length;
(6) if the chosen channel is larger than N, i.e. $(N, P]$, map these values to $[1, N]$ by modular operation;
(7) If the chosen channel is not available, replace it by a random available channel.

From the construction, each frame contains P numbers, each loop contains 3 frames, i.e. $3P$ numbers, and each round contains N loops, i.e. $3NP$ numbers. Therefore, the constructed sequence consists of N rounds with $3N^2P$ numbers. Figure 8.3 illustrates the construction (in [11], M represents the number of all channels). The construction can guarantee rendezvous between two (synchronous or asynchronous) (symmetric or asymmetric) users within $3N^2P$ time slots. We omit the details and reader may refer to [11] for details.

As shown in Table 8.1, although JS guarantees rendezvous between two symmetric users in a short time ($O(N)$ time slots), the *MTTR* value for two asymmetric users could be as large as $O(N^3)$ which is inacceptable.

Enhanced Jump Stay (EJS) [10] is a modified version such that rendezvous can be achieved in $O(N^2)$ time slots for two asymmetric users, but the main idea does not change. We would not introduce the modification and the readers who are interested in the JS algorithm could suggest some modifications that can reduce the rendezvous time.

8.5   Disjoint Relaxed Different Set (DRDS) Based Rendezvous Algorithm

We present an efficient rendezvous algorithm that has the best performance for both symmetric and asymmetric users [6]. The DRDS method guarantees rendezvous for two symmetric users in $O(N)$ time slots, and in $O(N^2)$ time slots for two asymmetric users. This method also can be modified such that two symmetric users can rendezvous in $O(1)$ time slots. We introduce the method in details in this section.

8.5.1   Global Sequence (GS)

We define the *Global Sequence (GS)* as follow:

Definition 8.1   We call $S = \{s_0, s_1, \ldots, s_{T-1}\}$ a Global Sequence (GS) where $\forall s_i \in S$, it is chosen from the full port set $U = \{1, 2, \ldots, N\}$.

Generally, the GS should contain every port (the label of the port) in U since the users are not aware of the available ports in U beforehand. We call the hopping sequence a *good GS* if the following two properties are satisfied.

Property 8.1   *The constructed GS has a fixed length T.*

Property 8.2   *For a GS $S' = \{s_0, s_1, \ldots, s_{T-1}\}$, $\forall \delta_t \geq 0$ and $\forall i \in C$, there exists $t \leq T$ such that $s_{t \bmod T} = i$ and $s_{(t+\delta_t) \bmod T} = i$.*

The first property guarantees that the users can repeat the sequence periodically since they can start the rendezvous process in different time slots. This is the main difference between two synchronous or asynchronous users. The second property guarantees rendezvous for any two port-asymmetric, asynchronous users once they begin to share some common available ports. Formally, we derive the result in the following theorem.

Theorem 8.1   *Two (port-symmetric or port-asymmetric, synchronous or asynchronous) users can rendezvous in T time slots if they both adopt a good GS of length T.*

Proof   Without loss of generality, supposing the available port sets for the two users u_i and u_j are C_i, C_j respectively, one user is $\delta \geq 0$ time slots earlier than the other; denote the good GS as $S = \{s_0, s_1, \ldots, s_{T-1}\}$ where $s_i \in U, 0 \leq i < T$.

For any common available port $k \in C_i \bigcap C_j$, according to Property 8.2, there exists $t \leq T$ such that:

$$\begin{cases} s_{t \bmod T} = k \\ s_{(t+\delta) \bmod T} = k \end{cases} \tag{8.7}$$

Since the user who starts earlier (no later than the other) accesses port $s_{t+\delta}$ while the other user accesses port s_t, rendezvous is achieved on port k in $t \leq T$ time slots, which concludes the theorem.

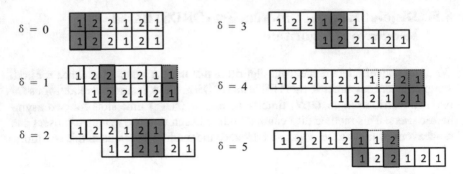

Fig. 8.4 An example of good GS

For example, we design a good GS for two ports as $\{1, 2, 2, 1, 2, 1\}$ which is of length 6; Fig. 8.4 shows that the sequence suits Property 8.2 since rendezvous is achieved on both ports $\{1, 2\}$ when one sequence is $0 \leq \delta < 6$ time slots earlier than the other.

Remark 8.1 Notice that any GS that guarantees rendezvous for two users should be a good GS, since the available ports are not known by the users beforehand and rendezvous has to be guaranteed on every port in $U = \{1, 2, \ldots, N\}$ no matter when the users start the rendezvous algorithm. Actually, Property 8.2 reveals the requirement to achieve rendezvous.

8.5.2 Disjoint Relaxed Difference Set (DRDS)

Before we show the method of constructing a good GS, we introduce some useful mathematical tools.

Relaxed difference set (RDS) is an efficient tool to construct cyclic quorum systems [8, 12]. We first introduce some definitions.

Definition 8.2 A set $D = \{a_1, a_2, \ldots, a_k\} \subseteq Z_n$ (the set of all nonnegative integers less than n) is called a Relaxed Difference Set (RDS) if for every $d \neq 0$ (mod n), there exists at least one ordered pair (a_i, a_j) such that $a_i - a_j \equiv d$ (mod n), where $a_i, a_j \in D$.

RDS is a variation of the (n, k, λ)-Difference Set [4, 12] where exactly λ ordered pairs (a_i, a_j) satisfying $a_i - a_j \equiv d$ (mod n) are required. Given any n, it is proved that any difference set D must have cardinality $|D| \geq \sqrt{n}$ [12]. The minimal D whose size approximates the lower bound can be found when $n = k^2 + k + 1$ and k is a prime power. Such a difference set is called a *Singer* Difference Set (SDS) [4]. For example, $D = \{1, 2, 4\}$ is both an SDS and an RDS under Z_7, but the set D is an RDS, not an SDS under Z_6. More information about difference sets can be found in the references and the readers may refer to them to find out more interesting properties.

In this section, we introduce one useful property which is the following.

Lemma 8.1 *If D is an RDS under Z_n, then $D_k = \{(a_i + k) \mod n | a_i \in D\}$ is also an RDS under Z_n.*

Proof From Definition 8.2, for every $d \neq 0 \pmod{n}$, there exists at least one ordered pair (a_i, a_j) where $a_i, a_j \in D$ satisfy $a_i - a_j \equiv d \pmod{n}$. Considering the set $D_k = \{(a_i + k) \mod n | a_i \in D\}$, denote $a_{k,i} = a_i + k, \forall a_i \in D$, then $D_k = \{a_{k,1}, a_{k,2}, \ldots, a_{k,|D|}\}$. For any $d \neq 0 \pmod{n}$, we choose $a_{k,i}, a_{k,j} \in D_k$ such that the corresponding values $a_i, a_j \in D$ satisfy $a_i - a_j \equiv d \pmod{n}$; then:

$$
\begin{aligned}
a_{k,i} - a_{k,j} &\equiv (a_i + k) - (a_j + k) \ mod \ n \\
&\equiv a_i - a_j \equiv d \ mod \ n
\end{aligned}
\tag{8.8}
$$

Therefore, the set D_k is also an RDS under Z_n from the definition.

Based on the definition of relaxed difference set, we introduce another important notation called *Disjoint Relaxed Difference Set*, as follows.

Definition 8.3 A set $S = \{D_1, D_2, \ldots, D_h\}$ is called a Disjoint Relaxed Difference Set (DRDS) under Z_n if $\forall D_i \in S$, D_i is an RDS under Z_n and $\forall D_i, D_j \in S, i \neq j$, $D_i \bigcap D_j = \emptyset$.

For example, $S = \{\{1, 2, 4\}, \{0, 3, 5\}\}$ is a DRDS under Z_6. Such a DRDS can be used to design GS based rendezvous algorithms and we will present the details later.

For any given integer n, there are many DRDSs under Z_n. Define Maximum DRDS S_n to be the set with the largest cardinality, and it is hard to find the maximum DRDS (see Lemma 8.5 in Sect. 8.6).

8.5.3   Equivalence of DRDS and Good GS

Before we present the method of achieving efficient rendezvous based on the introduced notations (DRDS and good GS), we first show their equivalence.

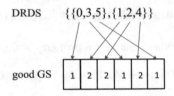

Fig. 8.5 An example of the equivalence between good GS and DRDS

Lemma 8.2 *Any DRDS corresponds to a* good *GS.*

Proof Consider a DRDS $S = \{D_0, D_1, \ldots, D_{h-1}\}$ under Z_n; we can construct a sequence $S' = \{s_0, s_1, \ldots, s_{n-1}\}$, as follows.

* If there exists D_j such that $i \in D_j$, let $s_i = j + 1$. Otherwise, assign any value in $[1, h]$ to s_i.

We claim that S' satisfies Properties 8.1–8.2. Obviously, Property 8.1 is satisfied since the sequence has length n. Then we show the satisfaction of Property 8.2. Without loss of generality, suppose one user starts δ_t time slots later. If $\delta_t \equiv 0$ (mod n), Property 8.2 is satisfied apparently. Let $d' = \delta + t \mod n$; thus $1 \leq d' < n$, and for any $i \in C$ where $C = \{1, 2, \ldots, h\}$, there exists a pair (a_j, a_k) where $a_j, a_k \in D_{i-1}$ and $a_j - a_k \equiv d'$ (mod n). Therefore, the property suits. Combining the two aspects, S' is a *good* GS.

Lemma 8.3 *Any* good *GS corresponds to a DRDS.*

Proof Consider a *good* GS $S' = \{s_0, s_1, \ldots, s_{T-1}\}$ on port set $C = \{1, 2, \ldots, N\}$; we construct the DRDS $S = \{D_0, D_1, \ldots, D_{N-1}\}$ under Z_T as follows:

* $D_i = \{j : s_j = i + 1, s_j \in S'\}$.

From Property 8.2, it is easy to check S is a DRDS.

For example, the DRDS $\{\{0, 3, 5\}, \{1, 2, 4\}\}$ corresponds to a good GS $\{1, 2, 2, 1, 2, 1\}$ as Fig. 8.5.

Based on Lemmas 8.2 and 8.3, we can construct a good GS for the users to achieve rendezvous if we can design the corresponding DRDS efficiently. Moreover, if there exists some efficient method to construct such good GS, we can solve some DRDS based problems.

8.5.4 DRDS Construction

To begin with, we present a DRDS construction method under Z_n in linear time where $n = 3P^2$ ($P > 3$ and P is a prime number).

Algorithm 8.1 DRDS Construction of Z_n when $n = 3P^2$

1: $S := \emptyset$;
2: **for** $i = 0$ to $P - 1$ **do**
3:     $D_i := (Z_{(3Pi+P)} \setminus Z_{3Pi})$;
4:     **for** $j = 0$ to $P - 1$ **do**
5:         $q_j := j^2$, $p_{ij} := \frac{(i-q_j)(P+1)}{2}$ mod P;
6:         $t_{j0} := 3Pj + P + p_{ij}$;
7:         $t_{j1} := 3Pj + 2P + p_{ij}$;
8:         $D_i := D_i \bigcup \{t_{j0}, t_{j1}\}$;
9:     **end for**
10:    $S := S \bigcup \{D_i\}$;
11: **end for**

Algorithm 1 constructs a DRDS $S = \{D_0, D_1, \ldots, D_{P-1}\}$ as follows: divide Z_n into P disjoint subsets

$$Z_n = U_0 \bigcup U_1 \bigcup \cdots \bigcup U_{P-1} \tag{8.9}$$

where $U_j = Z_{3P(j+1)} \setminus Z_{3P \cdot j}$. Let

$$D_i = T_{i0} \bigcup T_{i1} \bigcup \cdots \bigcup T_{i,P-1} \tag{8.10}$$

where $T_{ij} \subseteq U_j$.

For each U_j, let $q_j = j^2$ and $p_{ij} = \frac{(i-q_j)(P+1)}{2}$ mod P. Choose the $(P + p_{ij})$-th and $(2P + p_{ij})$-th number of U_j to compose T_{ij}. They are t_{j0} and t_{j1} (Lines 6,7). Then T_{ij} is constructed as:

$$\begin{cases} T_{ij} = \{t_{j0}, t_{j1}\} & \text{when } j \neq i \\ \{t_{j0}, t_{j1}\} \bigcup (Z_{(3Pi+P)} \setminus Z_{3Pi}) & \text{when } j = i \end{cases}$$

As the illustration in Fig. 8.6, Z_n is divided into P frames and each frame contains three segments of equal length. In constructing each set D_i, pick two numbers from the last two segments of each frame according to the above equations. In addition, all numbers in the first segment of the i-th frame are plugged into the set.

The intuitive idea of the construction is:

(1) In order to have some ordered pairs (a_j, a_k) satisfying $a_j - a_k \equiv d \pmod{n}$ when d is small from 1 to P, we choose the first P numbers in set U_i, i.e. $Z_{(3Pi+P)} \setminus Z_{3Pi}$;
(2) when d becomes much larger, we choose two numbers from each set U_j (the last two segments of each frame) at some appropriate positions according to the modular operations in Line 5.

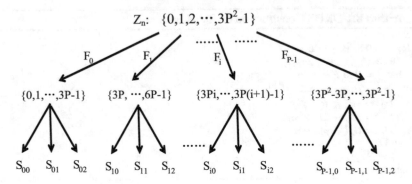

Fig. 8.6 Illustration of DRDS construction

We present a simple example when $n = 27$:

$$D_0 = \{0, 1, 2, 3, 6, 13, 16, 22, 25\};$$
$$D_1 = \{5, 8, 9, 10, 11, 12, 15, 21, 24\};$$
$$D_2 = \{4, 7, 14, 17, 18, 19, 20, 23, 26\}.$$

It is easy to verify that D_0, D_1, D_2 can compose a DRDS. We prove Algorithm 1 can indeed construct a DRDS formally.

Lemma 8.4 *Each set D_i constructed in Algorithm 8.1 is a RDS.*

Proof From the definition of RDS, we need to prove that: for any $d \neq 0 \,(\text{mod}\, n)$, there exists at least one ordered pair (a_j, a_k) satisfying $a_j - a_k \equiv d \,(\text{mod}\, n)$. Consider the following four cases:

(1) When $0 < d < P$: From Line 3 of Algorithm 1, P consecutive numbers are chosen, i.e. $3Pi, 3Pi + 1, \ldots, 3Pi + P - 1 \in D_i$; thus we can find $(3Pi + d, 3Pi)$ to meet the requirement;

(2) When $P \le d < 3P^2$ and $0 \le d \,(\text{mod}\, 3P) < P$: Assume $d = 3Pj_1 + b_1, 0 < j_1 < P, 0 \le b_1 < P$; we try to find one pair (a_j, a_k) such that

$$a_j = 3Pj_2 + b_2 \quad \text{mod}\, n$$
$$a_k = 3Pj_3 + b_3 \quad \text{mod}\, n$$

where $P \le b_3 < 2P$. If $a_j - a_k \equiv d \,(\text{mod}\, n)$, we can deduce $3Pj_2 + b_2 \equiv 3P(j_1 + j_3) + b_1 + b_3 \,(\text{mod}\, n)$, and thus $P \le b_2 < 3P$ and both b_2, b_3 satisfy the equality from Lines 6, 7 of Algorithm 1. Therefore:

$$b_2 \equiv \frac{(i - j_2^2)(P + 1)}{2} \quad \text{mod}\, P$$

$$b_3 \equiv \frac{(i - j_3^2)(P + 1)}{2} \quad \text{mod}\, P$$

Thus we have:

$$\begin{cases} j_2 \equiv (j_1 + j_3) & \mod P \\ \dfrac{(i - j_2^2)(P+1)}{2} \equiv \dfrac{(i - j_3^2)(P+1)}{2} + b_1 & \mod P \end{cases}$$

Combining these equations to derive:

$$2j_1 j_3 \equiv -(2b_1 + j_1^2) \quad \mod P \tag{8.11}$$

Since P is a prime number and j_1, b_1 are constant values when d is fixed, j_3 has one unique solution in Z_P [4] and we write the solution as j^*. Plugging j^* into the above equalities, we can compute the values of a_j and a_k.

For example, $P = 3, n = 27$, when $d = 11 = 3Pj_1 + b_1$, then $j_1 = 1, b_1 = 2$. Consider set D_1 and plug j_1, b_1 into Eq. (8.11):

$$2j_3 \equiv -5 \equiv 1 \quad \mod 3 \tag{8.12}$$

So $j_3 = 2$ and thus $j_2 = 0$, $b_3 \equiv 0 \pmod 3$. Since $3 \le b_3 < 6, b_3 = 3$ and then $b_2 = 5$. Therefore, $a_j = 3Pj_2 + b_2 = 5$ and $a_k = 3Pj_3 + b_3 = 21$. When $d = 11$, we can find such a pair $(5, 21)$ from D_1 to meet the requirement;

(3) When $P \le d < 3P^2$ and $P \le d \pmod{3P} < 2P$. Assume $d = 3Pj_1 + b_1, 0 \le j_1 < P, P \le b_1 < 2P$; let $c = \frac{(i-(i+j_1)^2)(P+1)}{2} \mod P$, $b = b_1 \mod P$ (both $c, b \in [0, P)$), and we find the pair (a_j, a_k) as:

$$a_j = \begin{cases} 3P(i + j_1) + P + c & \mod n \quad \text{if } c \ge b \\ 3P(i + j_1) + 2P + c & \mod n \quad \text{if } c < b \end{cases}$$

$$a_k = \begin{cases} 3Pi + c - b & \text{if } c \ge b \\ 3Pi + P + c - b & \text{if } c < b \end{cases}$$

It can be checked that $a_j, a_k \in D_i$ and $a_j - a_k \equiv d \pmod n$.

(4) When $P \le d < 3P^2$ and $2P \le d \pmod{3P} < 3P$. Assume $d = 3Pj_1 + b_1$, $0 \le j_1 < P, 2P \le b_1 < 3P$. Find (a_j, a_k) as in the second case:

$$a_j = 3Pj_2 + b_2 \quad \mod n$$
$$a_k = 3Pj_3 + b_3 \quad \mod n$$

The difference from the second case is $2P \le b_3 < 3P$; then $P \le b_2 < 3P$ and we can find out the appropriate j_2, j_3 values. Then apply the above equalities to derive a_j and a_k.

Based on the four cases above, $\forall d \ne 0 \pmod n$, we can find at least one ordered pair (a_j, a_k) such that $a_j - a_k \equiv d \pmod n$. Therefore, each set D_i constructed in the algorithm is a RDS.

Based on Lemma 8.4, we show that the constructed set of Algorithm 1 is a DRDS formally.

Theorem 8.2 *The set* $S = \{D_0, D_1, \ldots, D_{P-1}\}$ *constructed in Algorithm 1 is a DRDS.*

Proof From the definition of DRDS (Definition 8.3), we prove the theorem from two aspects:

(1) Each set $D_i \in S$ is an RDS;
(2) $\forall D_i, D_j \in S, i \neq j, D_i \bigcap D_j = \emptyset$.

From Lemma 8.4, we can check that each set $D_i \in S$ is an RDS. Then, we only need to prove that $\forall D_i, D_j \in S, i \neq j, D_i \bigcap D_j = \emptyset$.

From Algorithm 1:

$$D_i = T_{i0} \bigcup T_{i1} \bigcup \cdots \bigcup T_{i,P-1}$$
$$D_j = T_{j0} \bigcup T_{j1} \bigcup \cdots \bigcup T_{j,P-1}$$

It is clear that:

$$\forall k_1 \neq k_2, T_{i,k_1} \bigcap T_{j,k_2} = \emptyset \tag{8.13}$$

Therefore, we need to show:

$$\forall 0 \leq k < P, T_{ik} \bigcap T_{jk} = \emptyset \tag{8.14}$$

There are two situations:

(1) If $k \neq i, k \neq j$, two numbers from U_k are chosen for T_{ik}, T_{jk} respectively according to p_{ik} and p_{jk}. From Lines 6, 7 of Algorithm 1:

$$p_{ik} = \frac{(i - q_k)(P + 1)}{2} \mod P$$
$$p_{jk} = \frac{(j - q_k)(P + 1)}{2} \mod P$$

When $0 \leq i, j < P, i \neq j$ and we can conclude $p_{ik} \neq p_{jk}$. Thus $T_{ik} \bigcap T_{jk} = \emptyset$.
(2) If $k = i$ or $k = j$, the first P numbers of U_k will be chosen, while the other two numbers $3Pk + P + p_{ik}$ and $3Pk + 2P + p_{ik}$ do not intersect with the first P numbers, and thus $T_{ik} \bigcap T_{jk} = \emptyset$.

Combining in these two situations,

$$\forall k_1, k_2, T_{i,k_1} \bigcap T_{j,k_2} = \emptyset \tag{8.15}$$

and it implies

$$D_i \bigcap D_j = \emptyset \qquad (8.16)$$

Combining the two aspects, $S = \{D_0, D_1, \ldots, D_{P-1}\}$ is a DRDS.

It is obvious that Algorithm 8.1 constructs the DRDS with cardinality $\sqrt{\frac{n}{3}}$ and the algorithm runs in $O(n)$ time. The algorithm runs efficiently and it can be applied in designing efficient rendezvous algorithms.

8.5.5 DRDS Based Rendezvous Algorithm

Based on the DRDS construction of the special situation $n = 3P^2$ where P is a prime number,[1] we present the DRDS based rendezvous algorithm as follows.

Algorithm 8.2 DRDS Based Rendezvous Algorithm

1: Find the smallest prime P such that $P \geq N$;
2: **if** $P = 2$ **then**
3:   $T := 6, t := 0$;
4:   $S = \{D_0, D_1\}, D_0 = \{0, 1, 3\}, D_1 = \{2, 4, 5\}$;
5: **else**
6:   $T := 3P^2, t := 0$;
7:   Construct the DRDS $S = \{D_0, D_1, \ldots, D_{P-1}\}$ under Z_T as Algorithm 8.1;
8: **end if**
9: **while** Not rendezvous **do**
10:   **if** $0 \leq t < 2P$ **then**
11:     Access the port with smallest label in C';
12:   **else**
13:     $d := (t - 2P) \bmod T$;
14:     Find $D_i \in S$ such that $d \in D_i$;
15:     **if** Port $(i + 1) \in C'$ **then**
16:       Access port $(i + 1)$;
17:     **else**
18:       Access an available port in C' randomly;
19:     **end if**
20:   **end if**
21:   $t := t + 1$;
22: **end while**

Assume that the available port set for the user is $C' \subseteq U$ and the DRDS algorithm is described in Algorithm 8.2.

The first $2P$ time slots for the user is to access a fixed port, which resembles the listening period when a user wakes up in many asynchronous protocols; we call this the *Listening Stage*.

---

[1] *Bertrand-Chebyshev Theorem*: $\forall N > 1$, at least one prime P exists such that $N < P < 2N$.

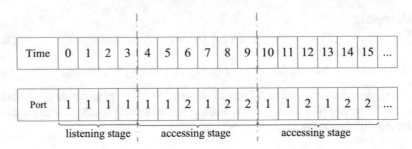

Fig. 8.7 An example of DRDS based rendezvous algorithm (Algorithm 8.2)

Afterwards, it is the *Accessing Stage* which repeats a GS of length $T = 3P^2$ based on the DRDS construction under Z_T from Algorithm 1 (if $P = 2$, the GS length is 6 and the DRDS is given as Line 2). Given any time t, compute $d = (t - 2P) \bmod T$ and find the RDS D_i that contains d. The user accesses port $(i + 1)$ if it is available; otherwise, it accesses a randomly picked available port.

Figure 8.7 is an example when $N = 2$ and $C' = \{1, 2\}$. The first four time slots form the *listening stage*, and in the *accessing stage*, the user repeats the sequence $\{1, 1, 2, 1, 2, 2\}$ of length $T = 6$.

We show the correctness and efficiency of Algorithm 8.2 formally.

Theorem 8.3 *For two users u_i and u_j with available port sets C_i, $C_j \subseteq U$, whenever they start Algorithm 8.2, rendezvous can be guaranteed within $MTTR = 3P = O(N)$ time slots if $C_i = C_j$, and $MTTR = 3P^2 + 2P = O(N^2)$ time slots if $C_i \neq C_j$.*

Proof It is easy to check that when $N \leq 2$, the theorem holds. For any $N \geq 3$, without loss of generality, suppose user u_i starts earlier at time 0 and user u_j starts at time $\delta_t \geq 0$. We derive the theorem from two situations:.

(1) If $C_i = C_j$, the best scenario for rendezvous is $0 \leq \delta_t < 2P$ because they are both in the *listening stage* accessing the same port. If $0 \leq (\delta_t - P) \pmod{3P} < 2P$, rendezvous occurs in the first $2P$ time slots when user u_j is *listening*, while user u_i is *accessing*. If $2P \leq (\delta_t - P) \pmod{3P} < 3P$, user u_j can achieve rendezvous in time $[2P, 3P)$ while keeping accessing some fixed port and the P numbers in $[\delta_t + 2P, \delta_t + 3P)$ for user u_i are in P different RDSs; so they achieve rendezvous in the *accessing stage*. Therefore, the maximum time to rendezvous is bounded in $3P$ time slots, i.e. $MTTR \leq 3P$.

(2) If $C_i \neq C_j$, we claim that rendezvous is guaranteed in $T + 2P$ time slots. Let $d = \delta_t \pmod{T}$. They may not achieve rendezvous in the *listening stage* even when $\delta_t < P$. For any common available port $i \in C_i \cap C_j$, we can find an ordered pair (a_j, a_k) from RDS D_{i-1} such that $a_j - a_k \equiv d \pmod{T}$ (Definition 8.2). So when user u_j's time ticks $a_k + P$, they both access port i, which implies rendezvous is guaranteed within $P + a_k \leq T + 2P$ time slots.

Combining the two aspects, the theorem holds.

8.5.6 Improved DRDS Based Rendezvous Algorithm

Although the DRDS based rendezvous algorithm (Algorithm 2) guarantees fast rendezvous for both symmetric and asymmetric users, we can improve it such that rendezvous can be achieved in $O(1)$ time slots for two symmetric users, which matches the state-of-the-art result [3] for the cognitive radio network.

Algorithm 8.3 Improved DRDS Based Rendezvous Algorithm

1: Find the smallest prime P such that $P \geq N$;
2: Denote the port with smallest label in C' as c_m and the label as m;
3: **if** $P = 2$ **then**
4:    $T_1 := 6, t := 0$;
5:    $S = \{D_0, D_1\}, D_0 = \{0, 1, 3\}, D_1 = \{2, 4, 5\}$;
6: **else**
7:    $T_1 := 3P^2, t := 0$;
8:    Construct the DRDS $S = \{D_0, D_1, \ldots, D_{P-1}\}$ under Z_T as Algorithm 8.1;
9: **end if**
10: $T := 6T_1$;
11: **while** Not rendezvous **do**
12:    $f := \lfloor t/6 \rfloor, d := t\%6$;
13:    **if** $d = 0$ or 1 or 3 **then**
14:      Find $D_i \in S$ such that $f \in D_i$;
15:    **else**
16:      $i := m - 1$;
17:    **end if**
18:    **if** Port $(i + 1) \in C'$ **then**
19:      Access port $(i + 1)$;
20:    **else**
21:      Access an available port in C' randomly;
22:    **end if**
23:    $t := t + 1$;
24: **end while**

Similar to Algorithm 8.2, assuming the available port set for the user is C' and denote the port with the smallest label as c_m where m is the label. Then Algorithm 8.3 constructs a DRDS S_1 similar to Algorithm 8.2. In order to guarantee fast rendezvous for two symmetric users, we expand each time slot into 6 slots where the 0, 1, 3-th numbers are the corresponding RDS in S (as Line 14) while the 2, 4, 5-th numbers are the smallest labels among all available ports. The time division method is introduced in Chap. 7. Figure 8.8 shows a simple example when there are only two available ports (we use m to be the smallest label in the figure other than 1).

We show the correctness and efficiency of Algorithm 8.3 as follows.

Theorem 8.4 *For two users u_i and u_j with available port sets $C_i, C_j \subseteq U$, whenever they start Algorithm 8.3, rendezvous can be guaranteed within $MTTR = 6 = O(1)$ time slots if $C_i = C_j$, and $MTTR = 18P^2 = O(N^2)$ time slots if $C_i \neq C_j$.*

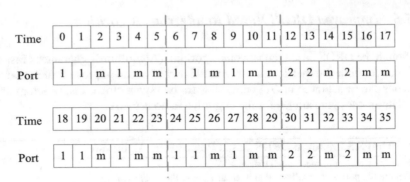

Time	0	1	2	3	4	5	6	7	8	9	10	11	12	13	14	15	16	17
Port	1	1	m	1	m	m	1	1	m	1	m	m	2	2	m	2	m	m

Time	18	19	20	21	22	23	24	25	26	27	28	29	30	31	32	33	34	35
Port	1	1	m	1	m	m	1	1	m	1	m	m	2	2	m	2	m	m

Fig. 8.8 An example of Improved DRDS based rendezvous algorithm (Algorithm 8.3)

Proof It is easy to check that when $N \leq 2$, the theorem holds. For any $N \geq 3$, without loss of generality, suppose user u_i starts earlier at time 0 and user u_j starts at time $\delta \geq 0$. Denote the smallest labels for both users as m_A, m_B respectively and we prove the theorem from two situations:

(1) If $C_i = C_j$, the smallest labels of the users are the same, i.e. $m_A = m_B$. Denote $x = \delta \% 6$ and there are six situations respectively.

  (1) If $x = 0$, both users access the port with the smallest label m_A, m_B at time $\delta + 2$, thus $TTR = 3$;
  (2) if $x = 1$, user u_i accesses port m_A at time $\delta + 4$ while user u_j accesses port m_B, thus $TTR = 5$;
  (3) if $x = 2$, user u_i accesses port m_A at time $\delta + 2$ while user u_j accesses port m_B, thus $TTR = 3$;
  (4) if $x = 3$, user u_i accesses port m_A at time $\delta + 2$ while user u_j accesses port m_B, thus $TTR = 3$;
  (5) if $x = 4$, user u_i accesses port m_A at time $\delta + 4$ while user u_j accesses port m_B, thus $TTR = 5$;
  (6) if $x = 5$, user u_i accesses port m_A at time $\delta + 5$ while user u_j accesses port m_B, thus $TTR = 6$;

  Thus rendezvous can be guaranteed in $MTTR = 6$ time slots.

  Actually, $\{\{0, 1, 3\}, \{2, 4, 5\}\}$ is a DRDS and the port with the smallest label occurs at the $2, 4, 5$-th time slots when we expand each time slot of the GS into 6 time slots, and these positions correspond the the second RDS. Therefore, $\forall x \in [0, 5]$, the two users' hopping sequences should intersect at some positions (of the expanded 6 time slots) so that both users access the port with the smallest label, and rendezvous can be guaranteed quickly;

(2) If $C_i \neq C_j$, similar to the first situation, we can conclude that two users' hopping sequences should intersect at some positions (of the expanded 6 time slots) so that both users access the ports corresponding to the RDS in S as in Line 14. From the proof of Theorem 8.3, $MTTR \leq 6 * 3P^2$ time slots.

Combining the two aspects, the theorem can be concluded.

Remark 8.2 In Chaps. 6 and 7, we present simple algorithms for two port-symmetric users if they are aware of the symmetric situation. However, they actually cannot know whether they are symmetric or not. The method introduced in this section is a good extension by transforming any rendezvous algorithm to a new one which can guarantee rendezvous in $O(1)$ time slots, if the two users are indeed port-symmetric. In addition, it would not degrade the time complexity as compared to the original algorithm if they are port-asymmetric.

Although the improved DRDS method can guarantee rendezvous for two symmetric users in a very short time, it increases the time to rendezvous for two asymmetric users (by 6 times). In practical situations, the original DRDS algorithm (Algorithm 8.2) is more preferable.

8.6   Lower Bound for GS Based Rendezvous Algorithms

In order to show the efficiency of the DRDS based rendezvous algorithm, we derive a lower bound for any Global Sequence (GS) based algorithm for two users. In other words, we should find the smallest length of any good GS based on N external ports, and any GS that guarantees rendezvous is a good GS (see Remark 8.1).

Since any good GS corresponds to a DRDS, one intuitive method is to find the DRDS with maximum cardinality under Z_n (denote the corresponding DRDS as maximum DRDS).

Lemma 8.5 *Given n, the cardinality of the maximum DRDS under Z_n is bounded by $|S_n| \le \sqrt{n}$.*

This lemma is derived easily from the fact that any RDS D should have cardinality $|D| \ge \sqrt{n}$ [12].

Actually, if there exists some algorithm $\mathscr{F}$ that can compute the DRDS of maximum cardinality under Z_n for any given $n > 0$, we can come up with an algorithm to derive the smallest length of good GS based on N channels as follows:

1: Invoke $\mathscr{F}$ to compute the maximum cardinality of any DRDS under Z_n where $n \in [N^2, 3P^2]$, where $P \ge N$ is a prime number;
2: Find the smallest n in the range such that the maximum cardinality is no less than N.

Here the smallest n is the smallest length of a good GS, i.e. the lower bound. In the first step, we try to compute the maximum cardinality of any DRDS under Z_n when n ranges from N^2 to $3P^2$. The value N^2 comes from Lemma 8.5, and the second value $3P^2$ comes from the DRDS based algorithm in this chapter. Therefore, the lower bound of the good GS can be derived precisely if we can compute the DRDS

Table 8.2 Relationship between n and maximum DRDS $|S_n|$ when $2 \leq n \leq 50$

The number: n	Maximum DRDS: $|S_n|$
$2 \leq n \leq 5$	1
$6 \leq n \leq 14$	2
$15 \leq n \leq 23$	3
$24 \leq n \leq 30$	4
31	5
$32 \leq n \leq 34$	4
$35 \leq n \leq 47$	5
$48 \leq n \leq 50$	6

with maximum cardinality. However, it is hard to compute the maximum DRDS for any given n, i.e. it is hard to find the tight bound (Lemma 8.5 is a loose bound).

Actually, for any set $\mathscr{D} = \{D_0, D_1, \ldots, D_h\}$ where D_i is an RDS under Z_n and $h \geq \sqrt{n}$, it is hard to compute the maximum DRDS from $\mathscr{D}$ since it can be reduced from the Set Packing Problem[2] which is NP-complete [9]. When each set $|D_i| \geq \sqrt{n}$, it is equivalent to Maximum $\sqrt{n}$-Set Packing which cannot be efficiently approximated within a factor of $\Omega(\frac{\sqrt{n}}{\ln \sqrt{n}})$ [7].

We compute all RDSs with cardinality in $[\sqrt{n}, \sqrt{3n}]$ and use exhaustive search to find the maximum DRDS when $n = 2, 3, \ldots, 50$. The relationship between n and the maximum DRDS (denoted as $|S_n|$) is listed in Table 8.2.

Since it is hard to compute the exact lower bound of any good GS, we try then derive a (loose) lower bound for any good GS based on the equivalence of DRDS and good GS. We first introduce an important lemma.

Lemma 8.6 *Suppose D is an RDS under Z_T where $T = N(N + 1)$ and $|D| = N + 1$, then $N \leq 3$.*

Proof Consider all pairs (a_j, a_k) where $a_j, a_k \in D$, $j \neq k$, and define $d_{jk} = (a_j - a_k) \bmod T$ which we call a *difference value*. $\forall d \in \{1, 2, \ldots, T - 1\}$, there exists at least one difference value $d_{jk} = d$. Since there are $N(N + 1)$ difference values, there exist two pairs (a_j, a_k) and (a'_j, a'_k) such that $d_{jk} = d_{j'k'}$ and the other difference values are all distinct. However,

$$d_{kj} = T - d_{jk} = T - d_{j'k'} = d_{k'j'} \tag{8.17}$$

which implies there exists another two pairs (a_k, a_j), (a'_k, a'_j) sharing a common difference value. The situation can happen only when $a_j = a'_k$, $a_k = a'_j$. Then

$$a_j - a_k \equiv a_k - a_j \quad \bmod T \tag{8.18}$$

and it means

---

[2]Given a finite set U and a list of subsets of U, the problem asks if some k subsets in the list are pairwise disjoint.

$$a_j - a_k \equiv \frac{T}{2} \quad \mod T \tag{8.19}$$

By Lemma 8.1, construct another RDS

$$D' = \{(a - a_j) \quad \mod T | a \in D\} \tag{8.20}$$

and thus $0, \frac{T}{2} \in D'$.

Denote

$$S_1 = \left\{ 0 < a < \frac{T}{2} | a \in D' \right\}$$

$$S_2 = \left\{ \frac{T}{2} < a < T | a \in D' \right\}$$

and let $d_1 = |S_1|$ and $d_2 = |S_2|$. Thus $d_1 + d_2 = |D| - 2 = N - 1$.

We count the number in set

$$S_3 = \left\{ 0 < a < \frac{T}{2} | a \notin D' \right\} \tag{8.21}$$

from two sides. First, since

$$|S_1 \bigcup S_3| = \frac{T}{2} - 1 \tag{8.22}$$

It is easy to compute:

$$d_1 + |S_3| = \frac{T}{2} - 1 \Rightarrow |S_3| = \frac{T}{2} - 1 - d_1 \tag{8.23}$$

From the analysis above, all other pairs satisfy:

$$(a_j, a_k) \neq \left(0, \frac{T}{2} \right) \text{ or } \left(\frac{T}{2}, 0 \right) \tag{8.24}$$

and hence it should have a distinct difference value. We construct S_3 as follows:

(1) $\forall a \in S_1$, let $a' = \frac{T}{2} - a \in S_3$; otherwise $(a, 0)$ and $(\frac{T}{2}, a')$ share the same difference value;
(2) $\forall a \in S_2$, let $T - a \in S_3$ and $a - \frac{T}{2} \in S_3$;
(3) $\forall a_1 < a_2 \in S_1$, define $\delta = a_2 - a_1$, and let $\frac{T}{2} - \delta \in S_3$ and $\delta \in S_3$; otherwise we can find two pairs sharing a common difference value;
(4) $\forall a_1 < a_2 \in S_2$, define $\delta = a_2 - a_1, 0 < \delta < \frac{T}{2}$, and then let $\frac{T}{2} - \delta$ and δ belong to S_3.
(5) $\forall a_1 \in S_1, \forall a_2 \in S_2$, define $\delta = a_2 - a_1$, if $\delta > \frac{T}{2}$; rewrite $\delta = T - \delta$, and then let $\delta \in S_3$ and $(\frac{T}{2} - \delta) \in S_3$.

It is easy to verify that when we choose one value a or two values $\{a, \frac{T}{2} - a\}$ to compose S_3, they cannot belong to S_3 before the step. (If $a = \frac{T}{2} - a$, we only add the value once and this special situation happens at most once.) Thus:

$$|S_3| \geq d_1 + 2d_2 + 2 \cdot \frac{d_1(d_1 - 1)}{2} + 2 \cdot \frac{d_2(d_2 - 1)}{2} + 2d_1 d_2 - 1 \qquad (8.25)$$

So:

$$\frac{T}{2} - 1 - d_1 \geq (d_1 + d_2)^2 + d_2 - 1 \qquad (8.26)$$

Plugging $d_1 + d_2 = N - 1$, we derive:

$$N^2 \leq 3N \Rightarrow N \leq 3 \qquad (8.27)$$

Therefore, the lemma holds.

Then, we derive a lower bound (not tight) for any good GS in Theorem 8.5.

Theorem 8.5 *Any good GS $S' = \{s_0, s_1, \ldots, s_{T-1}\}$ based on N channels satisfies:*

$$\begin{cases} T \geq N^2 + N & \textit{If } N \leq 2 \\ N^2 + N + 1 & \textit{If } N \geq 3 \textit{ and } N \textit{ is a prime power} \\ N^2 + 2N & \textit{Otherwise} \end{cases}$$

Proof When $N = 1$, it is clear that $T \geq 2$. Suppose $N \geq 2$; by Lemma 8.3, we can construct a DRDS as:

$$S = \{D_0, D_1, \ldots, D_{N-1}\} \qquad (8.28)$$

under Z_T. By Lemma 8.5, we have:

$$N \leq \sqrt{T} \Rightarrow T \geq N^2 \qquad (8.29)$$

Let $h = \min_{D_i \in S} |D_i|$; if $h \leq N$, the set D_i (where $|D_i| = h$) has exactly $h(h-1)$ ordered pairs (a_j, a_k), which implies at most $h(h-1) \leq N(N-1)$ difference values for d exist such that

$$a_j - a_k \equiv d \mod T \qquad (8.30)$$

When $N \geq 2$, we have:

$$N(N-1) < N^2 - 1 \leq T - 1 \qquad (8.31)$$

and D_i cannot be an RDS. Thus $h \geq N + 1$.
Assume $h = N + 1$, since

$$D_0 \bigcup D_1 \bigcup \cdots \bigcup D_{N-1} \subseteq Z_T \qquad (8.32)$$

we derive:

$$T \geq \sum_{i=0}^{N-1} |D_i| \geq Nh = N(N+1) \tag{8.33}$$

There are three cases to be analyzed.

Case 1: If $T = N(N+1)$, by Lemma 8.6, $N \leq 3$. When $N = 2$, $\{\{0, 1, 3\}, \{2, 4, 5\}\}$ is a DRDS under Z_6. However, when $N = 3$, we cannot find a DRDS with three disjoint RDS through exhaustive search;

Case 2: If $T = N^2 + N + 1$, suppose D_i suits $|D_i| = h$; since $(N+1)N = T - 1$, D_i is a $(T, h, 1)$-Difference Set. In [4], this is called a *Singer* Difference Set and it can be constructed only when N is a prime power. Thus when $N \geq 3$ and N is a prime power, $T \geq N^2 + N + 1$;

Case 3: If $T \geq N^2 + N + 2$ and N is not a prime power, suppose an RDS D_i suits $|D_i| = h$. It is clear that there are at most $h(h-1)$ ordered pairs (a_j, a_k) and the difference values $a_j - a_k \equiv d \pmod{n}$ cannot cover $\{1, 2, \dots, T-1\}$ since $h(h-1) = N(N+1) < N^2 + N + 1 \leq T - 1$, which implies D_i is not an RDS under Z_n, and so $h \geq N + 2$. From

$$D_0 \bigcup D_1 \bigcup \cdots \bigcup D_{N-1} \subseteq Z_T \tag{8.34}$$

we can conclude

$$T \geq \sum_{i=0}^{N-1} |D_i| \geq Nh \geq N(N+2) \tag{8.35}$$

Therefore, the theorem holds.

The lower bound is not always tight. Finding the minimum *good* CCHS length is (almost) equivalent to finding the maximum DRDS. As discussed above, it is hard to find the maximum DRDS, and thus it is also hard to find the tight lower bound for *good* CCHS. From Table 8.2, the lower bound of Theorem 8.5 is tight when $N = 1, 2, 5, 6$. However, when $N = 3, 4$, the lower bound for T is $13, 21$ respectively from the theorem, but the maximum DRDS $|S_n| = 2, 3$ under Z_n, implying the lower bound is not always tight.

Corollary 8.1 *Any GS based rendezvous algorithm cannot guarantee rendezvous in less than T time slots, where T is the expression in Theorem 8.5.*

Corollary 8.2 *The DRDS based algorithm (Algorithm 8.2) can achieve constant approximation as compared with the lower bound of any GS based blind rendezvous algorithms. Thus, it is a nearly optimal asynchronous rendezvous algorithm.*

Remark 8.3 We have not found a general method to construct a DRDS S under any Z_n such that $|S|$ is comparable to the bound in Lemma 8.5. However, if there exists such DRDS construction for arbitrary Z_n, we can transform it to a good rendezvous algorithm as shown in Lemma 8.2.

8.7   Chapter Summary

In this chapter, we study the blind rendezvous problem for two users by designing a good *global sequence (GS)* that is independent of the user's set of available ports. The intuitive idea is to hop through the external ports by repeating the GS for the users, and rendezvous can be guaranteed on some common available port if the GS satisfies some good properties.

In order to design a good GS for the users, we introduce an efficient mathematical tool called Disjoint Relaxed Difference Set (DRDS) which is shown to be equivalent to a good GS. Therefore, every algorithm that constructs a DRDS under the set Z_n in an efficient way can be adopted to construct a GS (of length n). In the chapter, we present a special construction of the DRDS under Z_n when $n = 3P^2$ where P is a prime number, and a good GS of length $3P^2 = O(N^2)$ can be reconstructed correspondingly, where N is the number of all ports and $P \geq N$ is a prime number. Therefore, blind rendezvous between two users is guaranteed in $O(N^2)$ time slots, which is the state-of-the-art result for GS based method.

Since the users in the network can start the rendezvous process freely and they may have different sets of available ports, the DRDS based rendezvous algorithm works under all the situations:

(1) *Port-Symmetric* and *Port-Asymmetric*: two symmetric users have the same set of available ports, while the sets for asymmetric users could be different;
(2) *Synchronous* and *Asynchronous*: two synchronous users start at the same time while asynchronous users are free to start the rendezvous process.

The DRDS based rendezvous algorithm guarantees fast rendezvous for two symmetric users by adding a listening stage, where the user accesses the available port with the smallest label for a sufficient long time. Moreover, the algorithm works for two asynchronous users because the GS satisfies the elegant property (Property 8.2) and it guarantees rendezvous in a very short time when the users are synchronous. The results for the four combinations are listed in Table 8.3.

Although the DRDS based rendezvous algorithm has good performance, we are eager to explore a general method to construct DRDS under Z_n where n is an arbitrary integer and to design a GS based algorithm which generates a good GS of length shorter than $3P^2$.

Table 8.3 *MTTR* values for the DRDS based rendezvous algorithm

DRDS	Symmetric	Asymmetric
Synchronous	1	$2P = O(N)$
Asynchronous	$3P = O(N)$	$P(3P - 1) = O(N^2)$

Remarks: 1) Improved DRDS algorithm guarantees rendezvous in $O(1)$ time slots for two symmetric users

References

1. Bian, K., Park, J.-M. & Chen, R. (2009). A quorum-based framework for establishing control channels in dynamic spectrum access networks. In *Mobicom*.
2. Bian, K., Park, J.-M., & Chen, R. (2011). Control channel establishment in cognitive radio networks using channel hopping. *IEEE Journal on Selected Areas in Communications, 29*(4), 689–703.
3. Chen, S., Russell, A., Samanta, A., & Sundaram, R. (2014). Deterministic blind rendezvous in cognitive radio networks. In *ICDCS*.
4. Colbourn, C. J., & Dintiz, J. H. (2006). *Handbook of Combinatorial Designs*. Boca Raton: CRC Press.
5. DaSilva, L., Guerreiro, I. (2008). Sequence-based rendezvous for dynamic spectrum access. In *DySPAN*.
6. Gu, Z., Hua, Q.-S., Wang, Y., Lau, F. C. M. (2013). Nearly optimal asynchronous blind rendezvous algorithm for cognitive radio networks. In *SECON*.
7. Hazan, E., Safra, S., & Schwartz, O. (2006). On the complexity of approximating k-set packing. *Computational Complexity, 15*(1), 20–39.
8. Jiang, J. R., Tseng, Y. C., & Lai, T. (2005). Quorum-based asynchronous power-saving protocols for IEEE 802.11 ad hoc network. *ACM Journal on Mobile Networks and Applications, 10*(1–2), 169–181.
9. Karp, R. M. (1972). Reducibility among combinatorial problems. *Complexity of Computer Computations*, 85–103.
10. Lin, Z., Liu, H., Chu, X., & Leung, Y.-W. (2013). Enhanced jump-stay rendezvous algorithm for cognitive radio networks. *IEEE Communications Letters, 17*(9), 1742–1745.
11. Liu, H., Lin, Z., Chu, X., & Leung, Y.-W. (2012). Jump-stay rendezvous algorithm for cognitive radio networks. *IEEE Transactions on Parallel and Distributed Systems, 23*(10), 1867–1881.
12. Luk, W. S., & Wong, T. T. (1997). Two new quorum based algorithms for distributed mutual exclusion. In *ICDCS*.
13. Shin, J., Yang, D., & Kim, C. (2010). A channel rendezvous scheme for cognitive radio networks. *IEEE Communications Letters, 14*(10), 954–956.
14. Theis, N. C., Thomas, R. W., & DaSilva, L. A. (2011). Rendezvous for cognitive radios. *IEEE Transactions on Mobile Computing, 10*(2), 216–227.
15. Yang, D., Shin, J., & Kim, C. (2010). Deterministic rendezvous scheme in multichannel access networks. *Electronics Letters, 46*(20), 1402–1404.

Chapter 9
Local Sequence (LS) Based Rendezvous Algorithms

Abstract In this chapter, we present *symmetric algorithms* for the blind rendezvous problem between two *asynchronous, non-anonymous* users. In the rendezvous setting, we fix *Alg*, *Time*, and *ID* as follows:

$$RS = < Alg\text{-}S, Asyn, Port, Non\text{-}Anon, Non\text{-}Obli > \qquad (9.1)$$

where $Port \in \{Port - S, Port - AS\}$, which implies that we design efficient algorithms that have good performance for both symmetric and asymmetric port situations. In designing Global Sequence (GS) based rendezvous algorithms for two anonymous users, many time slots are wasted since the user has to access a random available port if the pre-defined port in the sequence is not available. Since the users have no distinguishable identifiers, they have to obey the global sequence, which is verified to be an efficient method. However, if the users have identifiers (IDs), they are non-anonymous and they can decide on different hopping sequences based on its local information. GS based rendezvous algorithms construct a sequence of fixed length for all users and it is inefficient when the number of available ports accounts for only a small fraction of all external ports. Thus, we propose Local Sequence (LS) based rendezvous algorithms which construct different sequences on the basis of the local information for different users. In this chapter, we introduce LS based rendezvous algorithms where different users hop among their available ports according to different sequences, which are different from GS based algorithms in terms of intuition. By adopting the LS based rendezvous algorithms, the users may achieve rendezvous in a shorter time comparing with the GS based algorithms when the available ports are only a small fraction of all external ports. In Sect. 9.1, we give the motivation for constructing LS and present a simple example to illustrate. We first present the Ring Walk algorithm in Sect. 9.2, which constructs different sequences on the basis of the user's ID. Then, we present the Alternate Hop-and-Wait (AHW) algorithm in Sect. 9.3, which combines the hop and wait strategies. Two efficient LS based rendezvous algorithms are provided in Sects. 9.4 and 9.5 respectively, where the first one guarantees rendezvous in $O(N)$ and $O(N^2)$ time slots for two port-symmetric and port-asymmetric users respectively, and the latter guarantees rendezvous in $O(|C_i|)$ and $O((\max\{|C_i|, |C_j|\})^2)$ time slots for two port-symmetric and port-asymmetric users respectively, where C_i, C_j represent the available port sets of two users. Finally, we summarize the chapter in Sect. 9.6.

© Springer Nature Singapore Pte Ltd. 2017
Z. Gu et al., *Rendezvous in Distributed Systems*,
DOI 10.1007/978-981-10-3680-4_9

Time	0	1	2	3	4	5	6	7	8	9	10	11	...
sequence	1	1	1	1	3	2	1	3	2	2	2	2	...
User u_i	1	1	1	1	2	2	1	1	2	2	2	2	...
User u_j	-	2	3	2	3	3	2	2	3	2	2	2	...

Fig. 9.1 An example of global sequence based algorithm (DRDS [3])

9.1 Local Sequence (LS)

Although Global Sequence (GS) based algorithms can guarantee rendezvous for two users efficiently, much redundant information exists in achieving rendezvous. Technically, GS based algorithms construct a fixed sequence for all users and when the sequence comes to an unavailable port, the user has to replace it with a random available one or by some pre-defined rule. However, this (random) replacement always contains redundant information in rendezvous.

For example, let $U = \{1, 2, 3\}$, $C_i = \{1, 2\}$ and $C_j = \{2, 3\}$ for two users u_i and u_j respectively. Suppose user u_j is $\delta = 1$ time slot later than user u_i. As illustrated in Fig. 9.1, if both users run the GS based algorithm (for example, the DRDS based algorithm in Sect. 8.5.2) and user u_i replaces port 3 in the sequence by port 1 or 2 randomly, while user B replaces port 1 by port 2 or 3 (the replaced ports are labeled red in the figure). They can achieve rendezvous on the common port 2 in time slots 9, 10, 11. However, port 3 in the GS is useless for user u_i and port 1 is useless for user u_j. Thus we try to handle this problem by constructing different sequences for different users on the basis of local available ports, which we call *Local Sequence (LS)* based rendezvous algorithms.

In designing LS based rendezvous algorithms, we use the assumption in [2] that each user has a distinct identifier (ID). Different from [2], we assume the users' ID could be non-continuous in the range $[1, M]$, where M is the maximum value for the ID (the algorithm also works when M is a value larger than the number of users; in this book we use M for simplicity to denote the maximum ID value).

9.2 Ring Walk Algorithm

Ring Walk [5, 6] is the first LS based rendezvous algorithm for cognitive radio networks, where different sequences are constructed on the basis of the users' identifiers (IDs). The basic idea of Ring Walk is to represent each channel as a node in a ring and the construction of the hopping sequence is equivalent to visiting all these nodes in the ring.

Assume each user has a distinct ID in the range $[1, M]$ where M is the maximum value for the user's ID (in some settings, the maximum ID value can be larger than the number of users) and the users can generate hopping sequences by walking along the ring with a pre-assigned velocity, which corresponds to the value of the user's ID.

For example, the user with ID 1 generates the hopping sequence as:

$$\{1, 2, 3, \ldots, N, 1, 2, 3, \ldots, N, \ldots\} \tag{9.2}$$

and the user with ID $N - 1$ generates the hopping sequence as:

$$\{1, N, N - 1, N - 2, \ldots, 2, 1, N, N - 1, \ldots\} \tag{9.3}$$

Since different users have different velocities, i.e. different values of the IDs, they may meet in the ring according to the Chinese Remainder Theorem:

Theorem 9.1 Chinese Remainder Theorem: *For two numbers n_1, n_2 that are coprimes and two values a_1, a_2, there exists an integer x solving both equalities: (1) $x \equiv a_1 \pmod{n_1}$, (2) $x \equiv a_2 \pmod{n_2}$.*

This method can guarantee rendezvous between two asynchronous users, but it is inefficient, especially when the value of ID is large.

9.3 Alternate Hop-and-Wait (AHW) Algorithm

Alternate Hop-and-Wait (AHW)[2] also generates different sequences by assuming that each user has a unique ID. Different from the Ring Walk algorithm, AHW first scales the user's ID into a binary string of length $\log M$ and then designs different patterns for different bits. In the design of AHW, each pattern contains three modes and each mode contains P numbers. There are two types of modes: WAIT mode and HOP mode, where the WAIT model is similar to the stay frame of the JS algorithm (see Sect. 8.4), while the HOP mode is similar to the jump frame of the JS algorithm. By designing different patterns for different binary bits (there are three different bits including a special bit added to represent the starting symbol of the binary string), rendezvous can be guaranteed in $O(N^2 \log N)$ time slots (it is assumed that $M = N$ in [2]) for two asymmetric users, and in $O(N \log N)$ time slots for two symmetric users.

The AHW algorithm operates as follows. In the first place, the user's ID is represented by binary bits. For example, ID 10 is represented as $\{0, 0, 1, 0, 1, 0\}$ if we use 6 bits to represent the users' ID. Then, the algorithm adds a special bit 2 as the head of the string. Fig. 9.2 shows the construction.

Fig. 9.2 An example of modifying bit string

Original bits	0	0	1	0	1	0

Modified bits	2	0	0	1	0	1	0

Fig. 9.3 An example of
different patterns
corresponding to different
bits

Bit 0	WAIT	HOP	HOP

Bit 1	HOP	HOP	HOP

Bit 2	WAIT	WAIT	HOP

The two modes are constructed in a similar way to the JS algorithm, where the starting channel $i \in [1, N]$, and the step length $r \in [1, P)$ are utilized, where P is the smallest prime number larger than N. In the HOP mode, P different numbers are constructed:

$$a_j = (i + r * j - 1) \mod P + 1 \tag{9.4}$$

where a_j represents the j-th number. In the WAIT mode, the P numbers are the same as the starting channel i, i.e. $a_j = i$.

According to the bit strings of the user's ID, the AHW algorithm gives different patterns, as in Fig. 9.3. When the bit is 0, it corresponds to the three modes of WAIT HOP HOP; when the bit is 1, it corresponds HOP HOP HOP; when the bit is 2, it corresponds to WAIT WAIT HOP.

The AHW algorithm can be described formally as follows.

Step 1: Denote the binary strings of the user's ID as $I = \{i_1, i_2, \ldots, i_k\}$;
Step 2: Modify the binary string as $I' = \{2, i_1, i_2, \ldots, i_k\}$;
Step 3: Find the smallest prime number $P \geq N$; choose the starting channel as $i_0 \in [1, N]$ and the step length as $r \in [1, P)$;
Step 4: Construct the sequence with N rounds, where the starting channels of the j-th round is $i_j = (i_0 + i - 1) \mod N + 1$;
Step 5: In the j-th round, there are $k + 1$ frames and each frame contains 3 modes; on the basis of I', the 3 modes in each frame are constructed correspondingly (different patterns of modes, and the construction of HOP or WAIT mode is as above).

From the description, the constructed sequence contains $N * (k + 1) * 3P$ time slots, where k is the number of the bit strings that represent the user's ID. In each round, the starting channel will increase by 1 under modulo N while the step length remains the same. For example, for $N = 3$ channels and the user's ID that can be represented as $\{0, 1\}$, we construct the sequence as in Fig. 9.4.

Although both Ring Walk and AHW are LS based algorithms, they all use the labels of all licensed channels to construct the hopping sequences. In this chapter, we propose two distributed algorithms called Local Sequence (LS) based rendezvous and Modified Local Sequence (MLS) based rendezvous that only use the labels of the available channels and the user's ID. More details are provided in the following sections. Meanwhile, an elegant result has been proposed in [1] which constructs different sequences that are only based on the user's set of available channels (the user's ID is not utilized). The main technique adopted is graph coloring and the

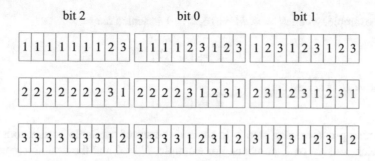

Fig. 9.4 An example of AHW algorithm

Table 9.1 MTTR Comparisons for LS based rendezvous algorithms

Algorithms	Symmetric	Asymmetric
Ring Walk [5]	$O(MN)$	$O(M^2N)$
AHW [2]	$3P \log M = O(N \log N)$	$3P^2 \log M = O(N^2 \log N)$
[1]	$O(1)$	$O(|C_A||C_B| \log \log N)$
LS [4]	$2(l+1)P = O(N)$	$2(l+1)P^2 = O(N^2)$
MLS [4]	$O(l_A|C_A|)$	$O(\max\{|C_A|, |C_B|\}^2)$

Remarks: (1) The comparisons are based on the rendezvous process between two users; (2) P is the smallest prime number $P \geq N$, $P = O(N)$; (3) C_A, C_B represent the sets of available channels for two users respectively; 4) l, l_A are some constants defined in Algorithm 9.1 and Algorithm 9.2 in Chap. 9

algorithm guarantees rendezvous in $MTTR = O(|C_A||C_B| \log \log N)$ time slots, where C_A, C_B represent the available channel sets for two users A and B respectively. There results are listed in Table 9.1.

9.4 A Simple LS Based Rendezvous Algorithm

In this section, we present a simple LS based rendezvous algorithm for a distributed system, which is different from both the Ring Walk algorithm and the AHW algorithm.

Suppose the user's identifier (ID) is $I \in [1, M]$ ($M = N^c$, c is a constant) and denote the available port set as $C' \subseteq U$. The intuitive idea of the method is to convert the user's ID to a certain fixed base number and construct different sequences based on the converted numbers (bits). Since different IDs have different representations, the constructed sequences could be different.

In the first place, the user's ID is scaled into $l = \lfloor \log_{P-1} M \rfloor + 1$ bits as shown in Algorithm 9.1, where P is the smallest prime number $P \geq N$. From the scaling steps, it is obvious that:

$$\forall i \in [0, l), 1 \leq d(i) < P \tag{9.5}$$

and different IDs have different scaled representations.

For example, when $N = 4$, $M = 16$, $I = 1$ is scaled as:

$$\vec{d} = \{1, 1, 2\} \tag{9.6}$$

while $I = 16$ is scaled as:

$$\vec{d} = \{2, 1, 1\} \tag{9.7}$$

Another preprocessing step is to expand the set of available ports into a vector $\vec{e}$, which consists of P numbers. For example, $N = 6$ and the available port set is $C' = \{2, 4, 5\}$, it is expanded as:

$$\vec{e} = \{2, 2, 2, 4, 5, 5, 5\} \tag{9.8}$$

Building on the preprocessing step, Algorithm 9.1 designs a sequence of length $T = 2(l + 1)P^2$ for the user. The construction can be thought of as constructing P periods of length $L = 2(l + 1)P$.

Algorithm 9.1 Local Sequence Based Algorithm

1: Find the smallest prime number $P \geq \max\{N, 3\}$;
2: $l := \lfloor \log_{P-1} M \rfloor + 1$;
3: ID Scale on I to get $\vec{d} = \{d(0), d(1), \cdots, d(l-1)\}$;
4: Expansion on C' to get $\vec{e} = \{e(0), e(1), \cdots, e(P-1)\}$;
5: $T := 2(l+1)P^2, t := 0, L := 2(l+1)P$;
6: **while** Not rendezvous **do**
7:    $t' := t \bmod T$;
8:    $x := \lfloor t'/L \rfloor, y := t' \bmod L$;
9:    **if** $y < 2P$ **then**
10:      $z := x$;
11:   **else**
12:      $y_1 := \lfloor (y - 2P)/(2P) \rfloor, y_2 := (y - 2P) \bmod 2P$;
13:      $z := (x + y_2 \cdot d(y_1)) \bmod P$;
14:   **end if**
15:   Access port $e(z)$;
16:   $t := t + 1$;
17: **end while**

ID Scale on I

1: **for** $i = l - 1$ to 0 **do**
2:    $d(i) := I \bmod (P - 1) + 1$;
3:    $I := \lfloor I/(P - 1) \rfloor$;
4: **end for**

Expansion on C'

1: Order the ports in C' as $c_1 < c_2 < \cdots < c_{|C'|}$;
2: Construct $\vec{e} = \{e(0), e(1), \cdots, e(P-1)\}$;
3: $e(j) := c_1, \forall 0 \leq j \leq c_2 - 2$;
4: **for** $i = 2$ to $|C'| - 1$ **do**
5:    $e(j) := c_i, \forall c_i - 1 \leq j \leq c_{i+1} - 2$;
6: **end for**
7: $e(j) := c_{|C'|}, \forall c_{|C'|} - 1 \leq j \leq P - 1$;

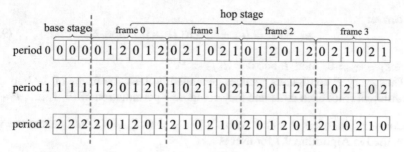

Fig. 9.5 An example of local sequence based algorithm

In each period, there is a base number x as in Line 8. For example the i-th period has base number $x = i$ and it stays the same for the first $2P$ time slots, which is the *base stage*. The following $2lP$ numbers are generated on the basis of the ID's scaled bits and it is the *hop stage*. The hop stage consists of l frames of length $2P$ and each frame relates to the scaled bit.

For example, the j-th frame is generated as

$$(i + k \cdot d(j)) \mod P, \forall 0 \le k < 2P \tag{9.9}$$

where $d(j)$ is called the *hopping step*. Then the corresponding port can be accessed as in Line 15 based on the expansion of set C'.

In order to guarantee rendezvous for two asynchronous users, each frame contains $2P$ numbers, which come from the idea of transforming non-aligned time slots into aligned ones (almost the same as Figs. 5.3 and 5.4). Moreover, the base stage is designed to accelerate the algorithm.

For example, for $N = 3$, $M = 9$, $I = 5$, the converted numbers are $\vec{d} = \{1, 2, 1, 2\}$ and three periods are constructed as in Fig. 9.5. Then the corresponding ports can be accessed according to the expansion of the available port set.

In order to derive the rendezvous time of the LS based algorithm, we first introduce an important lemma.

Lemma 9.1 *Every P continuous time slots in the same frame of the hop stage correspond to P different z values (Line 13 of Algorithm 9.1), i.e. these corresponding z values compose a permutation of $\{0, 1, \ldots, P - 1\}$.*

Proof Consider the j-th frame of period i, the $2P$ numbers are generated as:

$$z_k = i + k \cdot d(j) \mod P \forall 0 \le k < 2P \tag{9.10}$$

For any values $0 \le k_1, k_2 < 2P$ satisfying:

$$|k_1 - k_2| < P \tag{9.11}$$

we derive:

$$z_{k_1} - z_{k_2} = (k_1 - k_2) \cdot d(j) \neq 0 \quad \mod P \tag{9.12}$$

since $k_1 - k_2 \neq 0 \mod P$ and $0 < d(j) < P$.

Thus every P continuous z values generated in the same frame of the hop stage are different from each other and they compose a permutation of $\{0, 1, \ldots, P - 1\}$.

Consider two users A and B with $C_A \cap C_B \neq \emptyset$, $I_A \neq I_B$, and denote the variables used in Algorithm 9.1 for user A as:

$$(\overrightarrow{d_A}, \overrightarrow{e_A}, t_A, x(A), y_1(A), y_2(A)) \tag{9.13}$$

and the variables used for user B as:

$$(\overrightarrow{d_B}, \overrightarrow{e_B}, t_B, x(B), y_1(B), y_2(B)) \tag{9.14}$$

We derive the time complexity to achieve rendezvous for two symmetric users in the following Theorem.

Theorem 9.2 *LS algorithm (Algorithm 9.1) guarantees rendezvous in $MTTR = 2(l + 1)P = O(N)$ time slots for two port-symmetric users.*

Proof Two symmetric users (A and B) implies $C_A = C_B$, and thus $\overrightarrow{e_A} = \overrightarrow{e_B}$ can be verified easily. Since $I_A \neq I_B$, there exists $0 \leq i < l$ such that:

$$d_A(i) \neq d_B(i) \tag{9.15}$$

Without loss of generality, suppose user B is $\delta \geq 0$ time slot later than user A. We define:

$$\delta_T = \delta \quad \mod T \tag{9.16}$$

and define:

$$\delta_L = \delta \quad \mod L \tag{9.17}$$

According to different δ values, we show the theorem by examining six cases.

Case 1: $0 \leq \delta_L < 2P$ and $0 \leq \delta_T < 2P$. User B can achieve rendezvous with user A in the first time slot as shown in Fig. 9.6a, since user A is accessing port $e_A(0)$ in the base stage of Period 0 and user B's first attempt is $e_B(0) = e_A(0)$.

Case 2: $0 \leq \delta_L < P$ and $\delta_T \geq 2P$. Different from case 1, this situation means although user A is in the base stage when user B starts the algorithm, user A is accessing $e_A(k) \neq e_B(0), k > 0$, and thus they do not rendezvous during this stage.

Since there exists $0 \leq i < l$ such that $d_A(i) \neq d_B(i)$, they can achieve rendezvous in the i-th frame of the hop stage as Fig. 9.6b. Considering the time for user B suits:

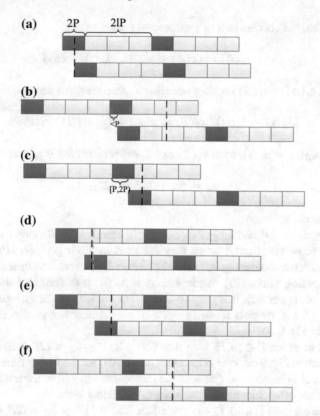

Fig. 9.6 Illustrations of Theorem 9.2's proof. The blocks labeled *gray* represent the base stages of each period and the *dotted lines* represent the situations that rendezvous happens

$$t_B \in [2(i+1)P, (2i+3)P) \tag{9.18}$$

from Line 8 and Line 12 of Algorithm 9.1, we can derive:

$$\begin{cases} x(B) = 0 \\ y_1(B) = i \\ y_2(B) \in [0, P) \end{cases}$$

The corresponding z values are generated as in Line 13:

$$z_t(B) = [0 + y_2(B) \cdot d_B(i)] \mod P \tag{9.19}$$

For user A, $t_A = t_B + \delta$, and from Line 8 and Line 12 of Algorithm 9.1, we derive:

$$\begin{cases} x(A) = \lfloor (t_B + \delta)/L \rfloor = \lfloor \delta_T/L \rfloor \\ y_1(A) = i \\ y_2(A) = y_2(B) + \delta_L \in [0, 2P) \end{cases}$$

Then, the corresponding z values are generated as:

$$z_t(A) = [x(A) + y_2(A) \cdot d_A(i)] \mod P \tag{9.20}$$

Let $z_t(A) = z_t(B)$ i.e. they access the same port, we can derive:

$$[d_B(i) - d_A(i)] \cdot y_2(B) = x(A) + \delta_L \cdot d_A(i) \mod P \tag{9.21}$$

As $d_B(i) \neq d_A(i)$, such $y_2(B)$ exists and rendezvous is guaranteed in time:

$$t_B \leq (2i + 3)P \leq (2l + 1)P \tag{9.22}$$

time slots;

Case 3: $P \leq \delta_L < 2P$ and $\delta_T \geq 2P$. Similar to case 2, user B cannot achieve rendezvous with user A when they are both in the hop stage. However, it is obvious that when $t_B \in [2P - \delta_L, 3P - \delta_L)$, user B is in the base stage accessing port $e_B(0)$, while user A is in the 0-th frame of the hop stage (in some period). It is easy to check that rendezvous can be guaranteed in $t_B \leq 2P$ time slots when $z_A = 0$ and user A accesses port $e_A(0) = e_B(0)$, as in Fig. 9.6c.

Case 4: There exists $i' \in [0, l)$ such that $(2i' + 2)P \leq \delta_L < (2i' + 3)P$. As illustrated in Fig. 9.6d, user B accesses port $e_B(0)$ for the first P time slots, while user A is in the same frame of the hop stage. Thus from the analysis of case 3, they can achieve rendezvous in $t_B \leq P$ time slots.

Case 5: There exists $i' \in [0, l - 1)$ such that $(2i' + 3)P \leq \delta_L < (2i' + 4)P$. Different from case 4, when $t_B \in [0, P)$, user A is not in the same frame, but rendezvous can be achieved when $t_B \in [P, 2P)$, as shown in Fig. 9.6e (the corresponding P time slots t_A are in the same frame).

Case 6: $2lP \leq \delta_L < (1 + 2l)P$. The situation is different from case 5 because when $t_B \in [P, 2P)$, user A is in the base stage and rendezvous may not happen. It is akin to case 2 that they can achieve rendezvous in the i-th frame where $d_A(i) \neq d_B(i)$ in $t_B \leq 2(l + 1)P$ time slots, as in Fig. 9.6f.

Combining the six situations, rendezvous for port-symmetric users can be achieved in $2(l + 1)P = O(N)$ time slots.

Similarly, we derive the time complexity to achieve rendezvous for two asymmetric users as follows.

Theorem 9.3 *LS algorithm (Algorithm 9.1) guarantees rendezvous in $MTTR = 2(l + 1)P^2 = O(N^2)$ time slots for two port-asymmetric users.*

Proof Since two users A and B are asymmetric, $C_A \neq C_B$, $I_A \neq I_B$. After the ID Scale and Expansion steps, the representations of the two users are different, i.e. we have different vectors:

$$\vec{d_A} \neq \vec{d_B} \tag{9.23}$$

and different expansions:

$$\vec{e_A} \neq \vec{e_B} \tag{9.24}$$

So there exists $0 \leq i < l$ such that:

$$d_A(i) \neq d_B(i) \tag{9.25}$$

As the two users share at least one common available port, there exists $0 \leq j < P$ such that:

$$e_A(j) = e_B(j) \tag{9.26}$$

The theorem can be proved based on the six cases in the symmetric scenario.

Case 1: $0 \leq \delta_L < 2P$ and $0 \leq \delta_T < 2P$. Different from symmetric users, $e_A(0)$ may not be equal to $e_B(0)$, and thus rendezvous is not guaranteed in the base stage of the 0-th period. However, when time counts to the j-th period, it is clear that when $t_B \in [2(l+1)P \cdot j, 2(l+1)P \cdot +P)$, user A and B are both in the base stage, accessing $e_A(j) = e_B(j)$. Thus $MTTR = t_B \leq 2(l+1)P^2$ time slots.

Case 2: $0 \leq \delta_L < P$ and $\delta_T \geq 2P$. When $t_B \in [k \cdot L + 2(i+1)P, k \cdot L + (2i+3)P)$, user B is in the i-th frame of the k-th period $(0 \leq k < P)$, and thus the corresponding $z_t(B)$ can be generated from Line 13:

$$z_t(B) = [k + y_2(B) \cdot d_B(i)] \mod P \tag{9.27}$$

From $t_A = t_B + \delta$, we can derive the $z_t(A)$ values as:

$$z_t(A) = [x(A) + k + y_2(A) \cdot d_A(i)] \mod P \tag{9.28}$$

where $x(A) = \lfloor \delta_T/L \rfloor$ is similar to case 2 of Theorem 9.2. Let $z_t(A) = z_t(B)$ to conclude the same result as Eq. (9.21). Denote:

$$\begin{cases} \theta = d_B(i) - d_A(i) \\ \lambda = x(A) + \delta_L \cdot d_A(i) \end{cases}$$

we can figure out:

$$y_2(B) = \lambda \cdot \theta^{-1} \tag{9.29}$$

where

$$\theta^{-1} \cdot \theta = 1 \mod P \tag{9.30}$$

Plugging $y_2(B)$ into Eq. (9.27), the corresponding $z_t(B)$ is computed. As k ranges in $[0, P)$, it is obvious that there exists $0 \leq k^* < P$ such that:

$$k^* + y_2(B) \cdot d_B(i) = j \mod P \tag{9.31}$$

which implies users A and B both access port $e_B(j) = e_A(j)$ at the same time. Thus rendezvous is guaranteed in $t_B \leq 2(l+1)P^2$ time slots.

Case 3: The other four cases discussed in Theorem 9.2 also can be proved in a similar way to case 1 or case 2. The readers who are interested in this part can derive the other four situations for a better understanding of the algorithm.

Combining these situations, we can conclude that rendezvous for two port-asymmetric users is bounded by $2(l+1)P^2 = O(N^2)$ time slots.

From Theorems 9.2 and 9.3, the LS based algorithm matches the best known results of Global Sequence based algorithms as shown in Table 8.1.

9.5   A Modified LS Based Rendezvous Algorithm

Although Algorithm 9.1 guarantees rendezvous for two users in a short time, it seems to be inefficient as the length of each frame is fixed as $2P$. When the number of available ports n is small, we could convert the specific user's ID to a new base number where the base and the length of each frame relate to n directly. The challenge is that different IDs may have the same representations in different base systems, such as $(12)_6 = 8$ but $(12)_4 = 6$ (here $(12)_6$ means 12 under base 6). Thus refinements should be made; we present the Modified Local Sequence (MLS) based algorithm as described in Algorithm 9.2.

Different from Algorithms 9.1, and 9.2 counts the number of available ports as $n = |C'|$ and finds the smallest prime number $p \geq n$. The preprocessing step of ID Scale is similar to Algorithm 9.1, but the ID is scaled by $p - 1$ where p relates to the number of available ports. Different users may have different p values, and thus the number of scaled bits for different users may be different since:

$$l := \lfloor \log_{p-1} M \rfloor + 1 \tag{9.32}$$

For example, $N = 5$, $M = 25$, suppose one user with ID $I = 5$ has $n = 3$ available ports, the ID is scaled as:

$$\vec{d} = \{1, 1, 2, 1, 2\} \tag{9.33}$$

For another user with ID $I = 5$ who has $n = 4$ available ports, the ID is scaled as:

$$\vec{d} = \{1, 2, 2\} \tag{9.34}$$

Clearly, the first user has 5 scaled bits while the second user has only 3 scaled bits.

Another preprocessing step is extraction on the available port set C', which is different from the expansion procedure in Algorithm 9.1. The extraction procedure constructs a vector $\vec{e}$ with p numbers by ordering the available ports as

---

Algorithm 9.2 Modified Local Sequence Based Algorithm

1: Count the number of available ports $n = |C'|$;
2: Find the smallest prime number $p \geq \max\{n, 3\}$;
3: $l := \lfloor \log_{p-1} M \rfloor + 1$;
4: ID Scale on I to get $\vec{d} = \{d(0), d(1), \cdots, d(l-1)\}$;
5: Extraction on C' to get $\vec{e} = \{e(0), e(1), \cdots, e(p-1)\}$;
6: $T := 2(l+1)p^2, t := 0, L := 2(l+1)p$;
7: **while** Not rendezvous **do**
8:    $t' := t \bmod T$;
9:    $x := \lfloor t'/L \rfloor, y := t' \bmod L$;
10:   **if** $y < 2p$ **then**
11:     $z := x$;
12:   **else**
13:     $y_1 := \lfloor (y - 2p)/(2p) \rfloor, y_2 := (y - 2p) \bmod 2p$;
14:     $z := (x + y_2 \cdot d(y_1)) \bmod p$;
15:   **end if**
16:   Access port $e(z)$;
17:   $t := t + 1$;
18: **end while**

ID Scale on I

1: **for** $i = l - 1$ to 0 **do**
2:   $d(i) := I \bmod (p - 1) + 1$;
3:   $I := \lfloor I/(p - 1) \rfloor$;
4: **end for**

Extraction on C'

1: Order the ports in C' as $c_1 < c_2 < \cdots < c_n$;
2: Construct $\vec{e} = \{e(0), e(1), \cdots, e(p-1)\}$;
3: **for** $j = 0$ to $p - 1$ **do**
4:   $i := j \bmod n + 1$;
5:   $e(i) := c_i$;
6: **end for**

$$c_1 < c_2 < \cdots < c_n \tag{9.35}$$

For example, for $N = 7$ and the available port set is $C' = \{1, 2, 4, 7\}$, the extraction result is:

$$\vec{e} = \{1, 2, 4, 7, 1\} \tag{9.36}$$

The number of vector $\vec{e}$ is related to the number of available ports n (actually, it is the smallest prime number that is no smaller than n), not all ports N.

Building on the preprocessing steps, Algorithm 9.2 constructs a sequence of length:

$$T = 2(l+1)p^2 \tag{9.37}$$

which also can be thought of as constructing p periods of length:

$$L = 2(l+1)p \tag{9.38}$$

Similar to Algorithm 9.1, there are also two stages in each period. The *base stage* consists of $2p$ base values $x = i$ for the i-th period as in Line 9, and the *hop stage* contains p frames. The $2p$ numbers of the j-th frame in the hop stage are generated as:

$$z = (i + k \cdot d(j)) \mod p, \forall 0 \le k < 2p \tag{9.39}$$

Then the corresponding port $e(z)$ is accessed as in Line 14. Algorithm 9.2 is a modified version of Algorithm 9.1, but it could be more efficient as the length of each user's sequence may be different. When the user has fewer available ports, the corresponding sequence is shorter.

Considering two users A and B with $C_A \cap C_B \ne \emptyset$ and $I_A \ne I_B$, denote the number of available ports for the two users as $n_A = |C_A|, n_B = |C_B|$ as in the first line of Algorithm 9.2. Similarly, denote the other variables used for user A in Algorithm 9.2 as:

$$(p_A, l_A, \overrightarrow{d_A}, \overrightarrow{e_A}, T_A, L_A, t_A, x(A), y_1(A), y_2(A), z_t(A)) \tag{9.40}$$

and the variables for user B as:

$$(p_B, l_B, \overrightarrow{d_B}, \overrightarrow{e_B}, T_B, L_B, t_B, x(B), y_1(B), y_2(B), z_t(B)) \tag{9.41}$$

We first derive the time complexity to achieve rendezvous for two symmetric users when they adopt the MLS algorithm.

Theorem 9.4 *The MLS algorithm (Algorithm 9.2) guarantees rendezvous in* $MTTR = 2(l_A + 1)p_A = O(l_A n_A)$ *time slots for two port-symmetric users.*

Proof Two symmetric users $(C_A = C_B)$ implies:

$$n_A = n_B, p_A = p_B, l_A = l_B, \overrightarrow{e_A} = \overrightarrow{e_B} \tag{9.42}$$

From the scaling step on ID, there exists $0 \le i < l_A$, such that:

$$d_A(i) \ne d_B(i) \tag{9.43}$$

The lengths of the two sequences are the same $(T_A = T_B)$ and from the proof details of Theorem 9.2, the theorem can be concluded similarly.

When the number of available ports is small, the MLS algorithm performs much better than the LS algorithm. It is clear that $l_A = O(\log N / \log n_A)$ and thus the $MTTR$ value could be small. For example, we derive the time complexity for different n_A values as:

$$n_A = O(1), MTTR = O(\log N)$$
$$n_A = O(\log N), MTTR = O(\log^2 N / \log \log N)$$
$$n_A = O(N^\varepsilon)(0 < \varepsilon < 1), MTTR = O(N^\varepsilon)$$

When it comes to asymmetric users, the situation is much more complicated. Before we derive the rendezvous time for two asymmetric users, we prove two important lemmas.

Lemma 9.2 *For two port-asymmetric users ($C_A \neq C_B$), rendezvous is guaranteed in $MTTR = 2(l_B + 1)p_B^2 = O(l_B n_B^2)$ time slots if $p_A = p_B$.*

Proof Two asymmetric users $C_A \neq C_B$ implies $\vec{e_A} \neq \vec{e_B}$. Since $p_A = p_B$, $I_A \neq I_B$, the number of scaled bits $l_A = l_B$, and there exists $0 \leq i < l_A$ such that $d_A(i) \neq d_B(i)$.

From $C_A \bigcap C_B \neq \emptyset$, there exist $0 \leq j_1, j_2 < p_A$ that suit:

$$e_A(j_1) = e_B(j_2) \tag{9.44}$$

The situation is similar to Theorem 9.3. For cases 1, 3, 4, 5 in the proof of Theorem 9.2, user B can achieve rendezvous in the base stage by accessing port $e_B(j_2)$ during:

$$t_B \in [2(1 + l_B)p_B \cdot j_2, 2(1 + l_B)p_B \cdot j_2 + 2p_B) \tag{9.45}$$

For the other two cases, rendezvous happens in the users' hop stage. The difference is: in Eqs. (9.27) and (9.28), we define:

$$\begin{cases} z_t(A) = j_1 \\ z_t(B) = j_2 \end{cases}$$

and it can be verified similarly that such $t_B < 2(l_B + 1)p_B^2$ exists. We omit the details and the readers can deduce the rendezvous time complexity according to the sketch.

Without loss of generality, suppose $p_B > p_A$; we derive the following lemmas.

Lemma 9.3 *Rendezvous is guaranteed within $MTTR = 2(l_B + 1)p_B^2 = O(l_B n_B^2)$ time slots if $p_B \geq 2p_A$.*

Proof Since $C_A \bigcap C_B \neq \emptyset$, we know:

$$\exists 0 \leq j_1 < p_A, 0 \leq j_2 < p_B, \ such \ that \ e_A(j_1) = e_B(j_2) \tag{9.46}$$

No matter which user starts the algorithm first, user B can achieve rendezvous in the base stage with period j_2. This is because the base stage contains $2p_B > 4p_A$ numbers, which is large enough to cover p_A continuous numbers from the same frame as user A's hop stage. Figure 9.7 illustrates the situation. Therefore, such $j_1 \in [0, p_A)$ exists and the $MTTR$ value is bounded by $2(1_B + 1)p_B^2$ time slots.

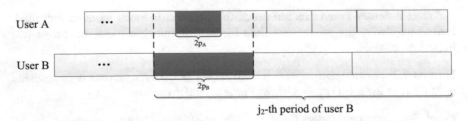

Fig. 9.7 An example of Lemma 9.3 when $p_B \geq 2p_A$. The blocks labeled *gray* represent the base stages of the users

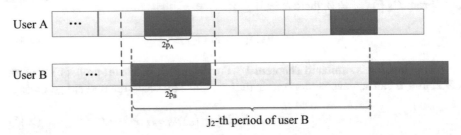

Fig. 9.8 An example of Lemma 9.4 when $p_A < p_B < 2p_A$. The blocks labeled *gray* represent the base stages of the users

Lemma 9.4 *Rendezvous is guaranteed within* $MTTR = 2(l_B + 1)p_B^2 p_A = O(l_B$ $n_B^2 n_A)$ *time slots if* $p_A < p_B < 2p_A$.

Proof Different from Lemma 9.3, as there are $2p_B$ numbers in the base stage, while there are $4p_A$ numbers in the hop stage of user A, any base stage of user B cannot cover a complete hop stage of user A. Thus, their rendezvous may not be guaranteed like that in Lemma 9.3. We analyze the worst situation for two users.

Since two users must have at least one common available port, we have:

$$\exists 0 \leq j_1 < p_A, 0 \leq j_2 < p_B, \text{ such that } e_A(j_1) = e_B(j_2) \tag{9.47}$$

Consider the base stage of the j_2-th period of user B, where:

$$t_B \in [\delta_B, \delta_B + 2p_B), \delta_B = 2(l_B + 1)p_B \cdot j_2 \tag{9.48}$$

Denote the corresponding time for user A as δ_A. As illustrated in Fig. 9.8, the only situation that user B cannot rendezvous in the base stage with user A is:

$$\begin{cases} L_A - p_A < (\delta_A \mod L_A) < L_A \\ 0 < (\delta_A + 2p_B \mod L_A) < p_A \end{cases} \tag{9.49}$$

Only when the two conditions are satisfied, user B may not achieve rendezvous in the base stage. Then user B repeats the sequence and we can determine how many time slots are needed to rendezvous. Denote:

$$\varepsilon = T_B \quad \mathrm{mod} \ L_A \tag{9.50}$$

It is clear that $\varepsilon \neq 0$. Only when $\varepsilon \in (0, p_A)$ or $\varepsilon \in (L_A - p_A, L_A)$, they may not rendezvous as user B repeats the sequence for the second time. However, if $\varepsilon \in (0, p_A)$, after at most $\frac{p_A}{\varepsilon}$ times, we have:

$$\left(\delta_A + \frac{p_A}{\varepsilon} \cdot T_B\right) \quad \mathrm{mod} \ L_A \in [0, P) \tag{9.51}$$

and rendezvous happens. If $\varepsilon \in (L_A - p_A, L_A)$, rendezvous is also guaranteed after $\frac{p_A}{L_A - \varepsilon}$ times. Thus $MTTR = 2(l_B + 1)p_B^2 p_A$ time slots and the lemma holds.

Lemma 9.4 reveals an extreme situation for the $MTTR$ value, which rarely happens. We show the $MTTR$ values on the basis of n_A, n_B for most cases in Tables 9.2–9.3. From Lemmas 9.2, 9.3 and 9.4, we have:

Theorem 9.5 *The MLS algorithm (Algorithm 9.2) guarantees rendezvous in $MTTR = O(l_B n_B^2)$ time slots if $p_B \geq 2p_A$ or $p_B = p_A$ and in $MTTR = O(l_B n_B^2 n_A)$ time slots if $p_A < p_B < 2p_A$.*

Combining Theorems 9.4 and 9.5, the MLS based algorithm is significantly better than the best known results in Table 8.1 when the number of available ports is small. Specifically, the MLS based algorithm can guarantee rendezvous in $O(l_A n_A)$ time slots for two symmetric users, which is much smaller than $O(N)$ when $n_A = o(N)$. It also guarantees rendezvous for two asymmetric users in less than $O(N^2)$ time slots for most combinations in Tables 9.2, 9.3.

Table 9.2 MTTR Comparisons with state-of-the-art rendezvous algorithms for different number of available ports (symmetric scenario)

n_A, n_B	$O(1)$	$O(\log N)$	$O(N^\varepsilon)$	$O(N)$
Algorithm				
JS [7]	$O(N)$	$O(N)$	$O(N)$	$O(N)$
DRDS [3]	$O(N)$	$O(N)$	$O(N)$	$O(N)$
AHW [2]	$O(N \log N)$	$O(N \log N)$	$O(N \log N)$	$O(N \log N)$
MLS [4]	$O(\log N)$	$O(\frac{\log^2 N}{\log \log N})$	$O(N^\varepsilon)$	$O(N)$

Remarks: (1) The comparisons are based on the rendezvous process between two symmetric users; (2) n_A and n_B represent the number of available ports for user A and B respectively; (3) ε is a constant in $(0, 1)$; 4)$n_A = n_B$ when they are symmetric

Table 9.3 MTTR Comparisons with state-of-the-art rendezvous algorithms for different number of available ports (asymmetric scenario)

n_B		$O(1)$	$O(\log N)$	$O(N^\varepsilon)$	$O(N)$
n_A					
$O(1)$	JS [7]	$O(N^3)$	$O(N^3)$	$O(N^3)$	$O(N^3)$
	DRDS [3]	$O(N^2)$	$O(N^2)$	$O(N^2)$	$O(N^2)$
	AHW [2]	$O(N \log N)$	$O(N \log_N^2)$	$O(N^{1+\varepsilon} \log N)$	$O(N^2 \log N)$
	MLS [4]	$O(\log N)$	$O(\frac{\log_N^3}{\log \log N})$	$O(N^{2\varepsilon})$	$O(N^2)$
$O(\log N)$	JS	$O(N^3)$	$O(N^3)$	$O(N^3)$	$O(N^3)$
	DRDS	$O(N^2)$	$O(N^2)$	$O(N^2)$	$O(N^2)$
	AHW	$O(N \log^2 N)$	$O(N \log_N^2)$	$O(N^{1+\varepsilon} \log N)$	$O(N^2 \log N)$
	MLS	$O(\frac{\log^3 N}{\log \log N})$	$O(\frac{\log^3 N}{\log \log N})$	$O(N^{2\varepsilon})$	$O(N^2)$
$O(N^\varepsilon)$	JS	$O(N^3)$	$O(N^3)$	$O(N^3)$	$O(N^3)$
	DRDS	$O(N^2)$	$O(N^2)$	$O(N^2)$	$O(N^2)$
	AHW	$O(N^{1+\varepsilon} \log N)$	$O(N^{1+\varepsilon} \log N)$	$O(N^{1+\varepsilon} \log N)$	$O(N^2 \log N)$
	MLS	$O(N^{2\varepsilon})$	$O(N^{2\varepsilon})$	$O(N^{2\varepsilon})$	$O(N^2)$
$O(N)$	JS	$O(N^3)$	$O(N^3)$	$O(N^3)$	$O(N^3)$
	DRDS	$O(N^2)$	$O(N^2)$	$O(N^2)$	$O(N^2)$
	AHW	$O(N^2 \log N)$	$O(N^2 \log N)$	$O(N^2 \log N)$	$O(N^2 \log N)$
	MLS	$O(N^2)$	$O(N^2)$	$O(N^2)$	$O(N^2)$

Remarks: (1) The comparisons are based on the rendezvous process between two asymmetric users; (2) n_A and n_B represent the number of available ports for users A and B respectively; (3) ε is a constant in $(0, 1)$

9.6 Chapter Summary

In this chapter, we study the blind rendezvous problem for two users and design *local sequence (LS)* based on local information: the identifier (ID) and the set of available ports. Although Global Sequence (GS) based algorithms can guarantee rendezvous in a short time, when the number of available ports accounts for only a small fraction of all the external ports, they are inefficient and many attempts have been wasted. Therefore, we propose LS based rendezvous algorithms to accelerate the rendezvous process, which work efficiently for both *symmetric* and *asymmetric* users, where two symmetric (or asymmetric) users have the same (or different) set of available ports.

Different from GS based rendezvous algorithms, each user is assumed to have a distinct ID to facilitate the designing of LS. *Channel hopping* and *ID scaling* are two important techniques used in the chapter (channel hopping is always used in designing rendezvous algorithms for cognitive radio networks), where channel hopping means the user hops through the ports with increasing labels by a fixed value which is called the hopping step, and ID scaling means the user's ID is scaled to some new number under another base and the scaled bits are used as the hopping steps in the LS.

In the chapter, several LS based rendezvous algorithms are introduced. The LS algorithm and the MLS algorithm work best among them. The LS algorithm scales the user's ID under base P where $P \geq N$ is a prime number, and then the hopping sequence is constructed with length $2(l+1)P^2 = O(N^2)$ where $l = \lfloor log_{P-1} M \rfloor + 1$ (in most situations, $l = O(1)$). Although the algorithm guarantees rendezvous for two symmetric users in $O(N)$ time slots and two asymmetric users in $O(N^2)$ time slots, it does not achieve significant improvement with respect to GS based algorithms. Therefore, we introduce a modified LS based rendezvous algorithm (MLS for short), which scales the user's ID under base p where $p \geq |C|$ (C is the user's set of available ports), and then the hopping sequence is constructed with length $2(l+1)p^2$ where $l = \lfloor \log_{p-1} M \rfloor + 1$ in a similar way. The MLS algorithm guarantees rendezvous for two symmetric users in $O(l_A |C_A|)$ time slots where l_A is the number of scaled bits, and for two asymmetric users in $O((\max\{|C_A|, |C_B|\})^2)$ time slots for most situations. Through comparison with the state-of-the-art GS based algorithms in Table 9.3, the MLS algorithm is shown to have better performance especially when the number of available ports is small.

References

1. Chen, S., Russell, A., Samanta, A., & Sundaram, R. (2014). Deterministic blind rendezvous in cognitive radio networks. In *ICDCS*.
2. Chuang, I., Wu, H.-Y., Lee, K.-R., & Kuo, Y.-H. (2013). Alternate hop-and-wait channel rendezvous method for cognitive radio networks. In *INFOCOM*.
3. Gu, Z., Hua, Q.-S., Wang, Y., & Lau, F. C. M. (2013). Nearly optimal asynchronous blind rendezvous algorithm for cognitive radio networks. In *SECON*.
4. Gu, Z., Hua, Q.-S., & Dai, W. (2014). Local sequence based rendezvous algorithms for cognitive radio networks. In *SECON*.
5. Lin, Z., Liu, H., Chu, X., & Leung, Y.-W. (2012). Ring-walk rendezvous algorithms for cognitive radio networks. *Ad Hoc and Sensor Wireless Networks*.
6. Liu, H., Lin, Z., Chu, X., & Leung, Y.-W. (2010). Ring-walk based channel-hopping algorithms with guaranteed rendezvous for cognitive radio networks. In *GreenCom-CPSCom*.
7. Liu, H., Lin, Z., Chu, X., & Leung, Y.-W. (2012). Jump-stay rendezvous algorithm for cognitive radio networks. *IEEE Transactions on Parallel and Distributed Systems, 23*(10), 1867–1881.

Chapter 10
Blind Rendezvous for Multi-users Multihop System

Abstract In the previous chapters of Part II, we introduce rendezvous algorithms between two users of different settings. As we know, typical distributed systems have a large number of entities, and two users wanting to communicate may not be able to that directly. Therefore, we extend the blind rendezvous algorithms between two users to multiple users in a multi-hop system in this chapter. Actually, the idea for the extension is not hard to follow. We take the Disjoint Relaxed Difference Set (DRDS) based rendezvous algorithm (please refer to Chap. 8) as an example. As described in Problem 5.2, the system consists of M users with available set C_i for user u_i and the common available port set $G = \bigcap_{i=1}^{m} C_i \neq \emptyset$. Suppose the diameter of the system is D, which implies any two users are connected within D hops. Notice that, if the system is not a connected component, no information can be exchanged between any disconnected components, and each separated disconnected component could run the rendezvous algorithm independently and the result would not affect the other components. We propose a distributed blind rendezvous algorithm that guarantees rendezvous for all users in $O(N^2 D)$ time slots in Sect. 10.1. The correctness and time complexity are derived in Sect. 10.2. We discuss about the rendezvous problem in a general multi-hop system in Sect. 10.3 and we summary the chapter in Sect. 10.4.

10.1 Algorithm Description

We adopt the intuitive idea in [1–3] to extend the DRDS algorithm for Problem 5.1 to the multiple users version: once every two users rendezvous on some common available port successfully, they can exchange the information about the set of available ports and they can synchronize their parameters (the set of available ports) such that they would generate the same hopping sequence afterwards.

Assume the available port set to be $C \subseteq U$, and we describe the algorithm as in Algorithm 10.1. The user runs the DRDS based rendezvous algorithm based on its own port set C. Once rendezvous happens with another user who has available port set $C' \subseteq U$, they can exchange their information about the available ports and update $C = C \bigcap C'$, and then continue the rendezvous process. It is easy to see that two users would generate the same hopping sequence until another rendezvous happens. The correctness and time complexity are analyzed in Theorem 10.1.

© Springer Nature Singapore Pte Ltd. 2017
Z. Gu et al., *Rendezvous in Distributed Systems*,
DOI 10.1007/978-981-10-3680-4_10

123

Algorithm 10.1 Blind Rendezvous Algorithm for Multiuser Multihop System

---

1: *Input:* $C \subseteq U$;
2: **while** Not terminated **do**
3:    Run the DRDS based rendezvous algorithm (Algorithm 8.2 in Chap. 8) based on available port set C;
4:    **if** Rendezvous with the user that has the available port set $C' \subseteq U$ **then**
5:       $C := C \cap C'$;
6:    **end if**
7: **end while**

10.2 Correctness and Complexity

Theorem 10.1 *Algorithm 10.1 guarantees that all users can achieve rendezvous in $MTTR = O(N^2 D)$ time slots, where D is the diameter of the network.*

Proof The theorem can be proved as follows. Since every two users u_i, u_j are connected within D hops, denote all the users along the shortest path connecting two users as:

$$\{u_{l_0}, u_{l_1}, u_{l_2}, \ldots, u_{l_k}\} \tag{10.1}$$

where $k \leq D$, $u_{l_0} = u_i$, $u_{l_k} = u_j$. We show that user u_{l_h} can update the set of available ports (denoted as C'_{l_h}) as a subset of $C_i \cap C_{l_h}$ in $O(N^2 h)$ time slots.

We apply the inductive method on l_h where $0 \leq h \leq k$:

(1) When $h = 0$, $u_i = u_{l_0}$ and thus the theorem holds;
(2) suppose when $h \leq h' < k$, user u_{l_h} can update the set of available ports as:

$$C'_{l_h} \subseteq C_i \cap C_{l_h} \tag{10.2}$$

in $O(N^2 h)$ time slots. Since user $u_{l_{h'+1}}$ can rendezvous with user $u_{l'_h}$ in $O(N^2)$ time slots and they would update (synchronize) their sets of available ports, thus we have:

$$C'_{l_{h'+1}} = C'_{l_h} \cap C_{l_{h'+1}} \subseteq C_i \cap C_{l_{h'+1}} \tag{10.3}$$

and the time is bounded by $O(N^2 (h + 1))$.

Combining the two cases, u_j can update the set of available ports as $C'_j \subseteq C_i \cap C_j$ in $O(N^2 D)$ time slots. Therefore, for any user u_j and any other user u_i in the network, the synchronized port set would be $C'_j \subseteq C_j \cap C_i$ after $O(N^2 D)$ time slots. Thus, the final available port set for user u_j should be $\bigcap_{i=1}^{m} C_i$ if all users have begun the rendezvous process for $O(N^2 D)$ time slots. Thus, all users hop through the available ports according to the same sequence generated by the DRDS based rendezvous algorithm. The theorem holds.

10.3 Discussions

In practice, the rendezvous process among multiple users in a multihop system could be more complicated. In this book, we make the assumption as in [3] that all users in the system share at least one common channel (the assumption is to suit cognitive radio networks, but should also apply to many distributed systems). This assumption is meaningful in some specific applications, such as file sharing where the system would dedicate a common external port, and message broadcasting which must be conducted on a common channel. A system is fully connected for communication if every two neighboring users share some common available port, which does not have be made known to *all* users.

Considering a Cognitive radio network where five SUs coexist with four PUs as depicted in Fig. 3.3 and they share the common available channel 5. When PU4 occupies channel 5, rendezvous between every pair of neighboring users can still happen. As illustrated in Fig. 10.1, (secondary) user A can rendezvous with user B on channel 5, user B can rendezvous with user C on channel 3, 4 or 5, user B can rendezvous with user D on channel 4 when channel 5 becomes unavailable for user D, and user D can rendezvous with user E on channel 1, 4 or 6. Although the five users do not share a single common available channel, rendezvous can happen between every pair of neighboring users, and thus the rendezvous process to cover the entire network can still be fulfilled.

This mode of communication takes place in many distributed systems, when the external ports are occupied by some unpredicted services or events. For example, in

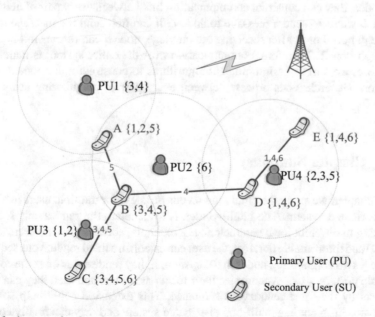

Fig. 10.1 An example of rendezvous among multiple users in a multihop CRN

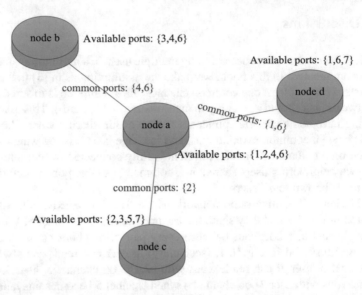

Fig. 10.2 An example of rendezvous among multiple entities in a distributed system

Fig. 10.2, node a can be connected to three neighbors and they have different sets of available ports. Node a and node b can achieve rendezvous on common port 4 or 6, node a and node d can rendezvous on common port 1 or 6, while node a and node c can only rendezvous on common port 2. Although there exists no common port among the four nodes, they can construct communication link between every pair of neighbors. If node b wants to send a message to node c, it can first send the message to node a through port 4 or 6; after receiving the message, node a can re-transmit it to node c through port 2. Therefore, communication over the entire system is maintained. Therefore, we aim to design efficient algorithms to construct a distributed system based on the rendezvous process between every pair of neighboring users in the future.

10.4   Chapter Summary

In this chapter, we study the blind rendezvous problem for multiple users in a multi-hop distributed system. The intuitive idea is to generate the rendezvous sequence according to different local parameters. For example, according to the users' identi-fier (ID) and the available ports set, the user can establish a fixed rendezvous sequence. For every two neighboring nodes in the system, if they rendezvous on some common available port, they can synchronize their local parameters and then they can access the ports by the same rendezvous sequence. This extension could help solve the rendezvous problem in a multi-hop distributed system and they can finally compute the common available port among all the users. This can be useful in some applications, such as one node tries to broadcast a message to all nodes through a common available port.

In many practical applications, there may not exist a common available port among all users but they can construct communication link between every two neighboring users. Therefore, it is a practical and good topic to design efficient rendezvous algorithms for a pair of neighboring users with short rendezvous time.

References

1. Chuang, I., Wu, H.-Y., Lee, K.-R., & Kuo, Y.-H. (2013). Alternate hop-and-wait channel rendezvous method for cognitive radio networks. In *INFOCOM*.
2. Gu, Z., Hua, Q.-S., Wang, Y., & Lau, F. C. M. (2013). Nearly optimal asynchronous blind rendezvous algorithm for cognitive radio networks. In *SECON*.
3. Liu, H., Lin, Z., Chu, X., & Leung, Y.-W. (2012). Jump-stay rendezvous algorithm for cognitive radio networks. *IEEE Transactions on Parallel and Distributed Systems*, *23*(10), 1867–1881.

Part III

Oblivious Blind Rendezvous in Distributed Systems

Chapter 11
Oblivious Blind Rendezvous

Abstract Time is divided into slots of equal length and each user can access an available channel in each time slot. Rendezvous is achieved only when the users access the same channel in the same time slot. All the extant blind rendezvous algorithms assume they know the global parameter N and the labels of these N channels, and some works [1] also assume each user knows the number of users in the network. In this part, we introduce the *oblivious blind rendezvous problem*, where *oblivious* means the entities' ports are labeled locally. As introduced in Part II, most blind rendezvous algorithms assume that all entities can see the same labels of the connected ports. However, this assumption is impractical in many distributed systems. For example, in cognitive radio networks, many works assume the licensed spectrum is divided into N non-overlapping channels with fixed labels $\{1, 2, \ldots, N\}$, and each user can access the channel not occupied by any nearby PUs as an *available channel*. However, this assumption may not align with the reality when designing blind rendezvous algorithms. Actually, all users may not see the same labels for the licensed channels. For example, the 'TV white space' that could be sensed by the users has operating frequencies ranging from 470–790 MHz in Europe [2, 4], but it is located in the VHF (i.e. very high frequency) (54–216 MHz) and UHF (i.e. ultra high frequency) (470–698 MHz) bands in the United States [3]. Obviously, the labeling of this space could be different and the same frequency band (channel) may be assigned different labels under different administrations. In a general distributed system, each user has N external ports and it can label these ports locally from $\{1, 2, \ldots, N\}$ in order to distinguish them. Any port k of user u_i may not be connected with port k of user u_j since both users may only use k to identify the different ports. In some special applications, the ports may be labeled according to a global rule. For example, the FTP service uses port 21 of the computers, and the default port for WWW service is 80. We study a more general situation where the users do not have a common labeling rule, and this can be used in many general applications. In this chapter, we first present the system model for the oblivious blind rendezvous problem, in Sect. 11.1; then we introduce the commonly used metrics for evaluation in Sect. 11.2. The problem definition is provided in Sect. 11.3 and we give examples of oblivious blind rendezvous for better understanding in Sect. 11.4. Finally, we summarize the chapter in Sect. 11.5.

© Springer Nature Singapore Pte Ltd. 2017
Z. Gu et al., *Rendezvous in Distributed Systems*,
DOI 10.1007/978-981-10-3680-4_11

11.1   System Model

In this part, we present the rendezvous algorithms for different types of rendezvous settings, on the basis that the ports are oblivious, i.e. all entities do not apply the same labeling rules. Therefore, the rendezvous settings can be represented as:

$$RS_{oblivous} =< Alg, Time, Port, ID, Obli > \qquad (11.1)$$

where $Alg \in \{Alg - AS, Alg - S\}$, $Time \in \{Syn, Asyn\}$, $Port \in \{Port - S,$ $Port - AS\}$, and $ID \in \{Non - Anon, Anon\}$.

Technically speaking, suppose there are $M(M \geq 2)$ users in a distributed system, and each user has $N(N \geq 1)$ external ports. In the $RS_{oblivious}$ setting, the ports of each user can be labeled freely by the user itself. Denote all users as:

$$\{u_1, u_2, \ldots, u_M\} \qquad (11.2)$$

For simplicity, we assume that each external port have a universal label which is not seen by the users. Denote the set of ports with universal labels as:

$$U = \{u_1, u_2, \ldots, u_N\} \qquad (11.3)$$

For any user u_i, suppose the user labels the N ports locally as:

$$\{p_i(1), p_i(2), \ldots, p_i(N)\} \qquad (11.4)$$

For any two users u_i, u_j, the ports $p_i(k)$ and $p_j(k)$ may not be connected after the local labeling.

Suppose the adopted rendezvous algorithms of the users are:

$$\{F_1, F_2, \ldots, F_M\} \qquad (11.5)$$

respectively (we suppose user u_i runs algorithm F_i).

(1) In the $Alg - AS$ setting, for any two users $u_i, u_j, i \neq j$, F_i and F_j could be different (we use $F_i \neq F_j$ to indicate that they are different).
(2) In the $Alg - S$ setting, all users share the same algorithm, i.e. $\forall i, j \in [1, M]$, $F_i = F_j$.

Similar to blind rendezvous in the distributed systems, time is also assumed to be divided into slots of equal length $2t$, where t is the sufficient time for establishing a communication link between two connected ports. Suppose the system is slot-aligned and each user can choose a port for rendezvous attempt in each time slot. If two users' time slots are not aligned, we can also transfer it to slot-aligned scenario as in Fig. 5.3 in Chap. 5.

Denote the start time of the users as:

$$\{t_1, t_2, \ldots, t_M\} \tag{11.6}$$

respectively, where the start time of user u_i is t_i.

(1) In the *Syn* setting, all users have the same start time, i.e. $\forall i, j \in [1, M], t_i = t_j$.
(2) In the *Asyn* setting, for any two users $u_i, u_j, i \neq j$, t_i and t_j could be different, i.e. $t_i \neq t_j$.

Similar to the blind rendezvous problem, some ports of each user may be occupied by other services, and the user could use only a fraction of the N external ports. We say a port is *available* if it is not occupied by others and the user can choose it for communication. For any user u_i, it can sense an available port set as $C_i \subseteq U$. Although user u_i may have already labelled the ports locally, it can also label the available ports as

$$C_i = \{c_i(1), c_i(2), \ldots, c_i(k_i)\} \tag{11.7}$$

where $k_i = |C_i|$ represents the number of available ports. Actually, we can regard each available port $c_i(j)$ as a port with global label u_l.

(1) In the *Port* − *S* setting, all users have the same *global* available ports, i.e. for each user u_j and user $u_j, k_i = k_j$, and $\forall l_i \in [1, k_i]$, there exists $l_j \in [1, k_j]$ such that port $c_i(l_i)$ and port $c_j(l_j)$ correspond to the same global label (they are connected).
(2) In the *Port* − *AS* setting, all users may not have the same available ports, i.e. for each user u_i and user $u_j, \exists l_i \in [1, k_i]$ such that port $c_i(l_i)$ is not connected to any available port in C_j.

In the *Port* − *AS* setting, in order to guarantee rendezvous, two neighboring users must have at least one common available port, i.e. for any two neighboring users u_i, u_j, there exists $l_i \in [1, k_i]$ and $l_j \in [1, k_j]$ such that $c_i(l_i)$ and $c_j(l_j)$ correspond to the same global port, which indicates they are connected. For simplicity, we denote $C_i \bigcap C_j \neq \emptyset$.

In designing oblivious blind rendezvous algorithms, the users' identifiers (IDs) play an important role. Therefore, we define the settings as follows:

(1) In the *Anon* setting, all users are anonymous and they have no distinct identifiers.
(2) In the *Non* − *Anon* setting, each user has a distinct identifier (ID). Denote user u_i's ID as I_i. For any two users $u_i, u_j, i \neq j$, I_i and I_j are different, i.e. $I_i \neq I_j$.

Actually, as there are M users in all and we suppose the a user's ID is a distinct number in the range $[1, \hat{M}]$, where $\hat{M}$ means the maximum ID value for the users. Some works assume $\hat{M} = M$, which means the user could only have continuous IDs in the range [1]. In this book, we assume $\hat{M}$ could be larger than M, but we assume it is bounded as $\hat{M} \leq N^c$ where c could be any arbitrary large constant. For simplicity, we re-use notation M to denote $\hat{M}$ in the following chapters.

11.2 Metrics

We use *Time to Rendezvous (TTR)* to measure the efficiency of rendezvous algo-rithms. As introduced in the model, the start time of user u_i is denoted as t_i. Suppose the finish time of user u_i is d_i, where $d_i > t_i$. For any two neighboring users u_i and u_j, both users will finish the process at the same time if they achieve rendezvous, thus $d_i = d_j$.

We define the time to rendezvous between two users in oblivious blind rendezvous as:

Definition 11.1 For two neighboring users u_i and u_j, suppose their start time are t_i, t_j respectively, and their finish time are d_i, d_j where $d_i = d_j = d$. The time to rendezvous is defined as:

$$TTR = d - \max\{t_i, t_j\} \tag{11.8}$$

We also denote the rendezvous time as the elapsed time of the user who starts the rendezvous process later. We define the time to rendezvous among all M users as:

Definition 11.2 Considering all user $\{u_1, u_2, \ldots, u_M\}$ in the system, denote their start time and finish time as $\{t_1, t_2, \ldots, t_M\}$ and $\{d_1, d_2, \ldots, d_M\}$ respectively. The time to rendezvous is defined as:

$$TTR = \max_{1 \leq i \leq M} d_i - \max_{1 \leq i \leq M} t_i \tag{11.9}$$

We also use two important metrics to evaluate the proposed rendezvous algo-rithms:

(1) *Maximum Time to Rendezvous (MTTR)* represents the maximum time used to rendezvous in all different situations, such as different available ports, different start times, etc.
(2) *Expected Time to Rendezvous (ETTR)* represents the expected time used to rendezvous in all different situations.

$MTTR$ reveals the performance of the rendezvous algorithm in the worst situa-tion, while $ETTR$ reveals the average performance.

11.3 Problem Definition

As described in the System Model, there are M users and their available ports can be denoted as:

$$\{C_1, C_2, \ldots, C_M\} \tag{11.10}$$

If rendezvous happens for all users, denote the common available port set as:

$$G = \bigcap_{i=1}^{M} C_i \neq \emptyset \tag{11.11}$$

which are the common available ports (with global labels). Before we define the rendezvous problem for multiple users in the system, we first formulate the *oblivious blind rendezvous (OBR)* problem between two users, as follows:

Problem 11.1 OBR-2: Given an available channel set $C \subseteq U$ and the ID $I \in [1, M]$, design an algorithm to access global ports over different time slots $t : f_{C,I}(t) \in C$ such that for any two users u_i and u_j with $C_i, C_j \subseteq U, C_i \bigcap C_j \neq \emptyset$ and ID $I_i, I_j \in [1, M], I_i \neq I_j$ respectively,

$$\forall \delta, \exists T_\delta, \; s.t. \; f^i_{C_i, I_i}(T_\delta + \delta) = f^j_{C_j, I_j}(T_\delta). \tag{11.12}$$

The TTR value is T_δ when user u_j starts the rendezvous process δ time slots later than user u_i. The $MTTR$ value of the algorithms is defined as:

$$MTTR_{f^i, f^j} = \max_{\forall \delta} T_\delta \tag{11.13}$$

The objective is to design rendezvous algorithms with bounded $MTTR$ value to guarantee rendezvous between two users. Notice that, f^i represents the algorithm user u_i adopts. If we are to design *symmetric algorithms* for the users, both users u_i and u_j should adopt the same algorithm, i.e. $f^i = f^j$.

Remark 11.1 When user u_j starts the rendezvous process earlier than user u_i, $\delta < 0$ in Eq. (11.12).

Based on the rendezvous problem definition of two users, we formulate the *Oblivious Blind Rendezvous Problem for Multiple Users in the Multihop system* as follows:

Problem 11.2 Consider a multihop system with $M(M \geq 2)$ users where each user has a distinct ID $I \in [1, M]$. Denote the available port set for user u_i as:

$$C_i = \{c_i(1), c_i(2), \ldots, c_i(k_i)\} \tag{11.14}$$

where $k_i = |C_i|$. Let $G = \bigcap_i C_i$ and $G \neq \emptyset$. Design distributed algorithms for the users such that all users are guaranteed to rendezvous on the same port in G, regardless of the different times when the users begin the process.

11.4   Examples of Oblivious Blind Rendezvous

Figure 11.1 is an example for OBR-2. Assume there are 5 external ports:

$$U = \{u_1, u_2, u_3, u_4, u_5\} \tag{11.15}$$

of which u_1, u_3 are available for user u_a with ID $I_a = 1$:

$$C_a = \{c_a(1), c_a(2)\} \tag{11.16}$$

and the ports are labeled locally as:

$$\begin{cases} c_a(1) = u_1 \\ c_a(2) = u_3 \end{cases}$$

Meanwhile, u_1, u_2, u_4, u_5 are available for user u_b with ID $I_b = 2$:

$$C_b = \{c_b(1), c_b(2), c_b(3), c_b(4)\} \tag{11.17}$$

and the ports are labeled locally as:

$$\begin{cases} c_b(1) = u_5 \\ c_b(2) = u_4 \\ c_b(3) = u_2 \\ c_b(4) = u_1 \end{cases}$$

Consider a simple algorithm: user u_a repeats accessing the ports according to the sequence:

$$\{c_a(1), c_a(1), c_a(2), c_a(2)\} \tag{11.18}$$

while user u_b accesses ports according to the sequence:

Time_a	1	2	3	4	5	6	7	8
user u_a Sequence	$c_a(1)$	$c_a(1)$	$c_a(2)$	$c_a(2)$	$c_a(1)$	$c_a(1)$	$c_a(2)$	$c_a(2)$
Port	u_1	u_1	u_3	u_3	u_1	u_1	u_3	u_3

	Time_b	1	2	3	4	5	6	7	8
user u_b	Sequence	$c_b(1)$	$c_b(2)$	$c_b(3)$	$c_b(4)$	$c_b(1)$	$c_b(2)$	$c_b(3)$	$c_b(4)$
	Port	u_5	u_2	u_4	u_1	u_5	u_2	u_4	u_1

Fig. 11.1   An example of OBR-2

Time$_a$	1	2	3	4	5	6	7	9	10
user u_a Sequence	1	2	1	2	1	2	1	2	1
Port	$c_a(1)$	$c_a(2)$	$c_a(1)$	$c_a(2)$	$c_a(1)$	$c_a(2)$	$c_a(1)$	$c_a(2)$	$c_a(1)$
user u_b Sequence	-	1	2	3	4	1	2	3	4
Port	-	$c_b(1)$	$c_b(2)$	$c_b(3)$	$c_b(4)$	$c_b(1)$	$c_b(2)$	$c_b(3)$	$c_b(4)$

Fig. 11.2 An example of OBR-2 when the users adopt a symmetric algorithm and $\delta = 1$

Time$_a$	1	2	3	4	5	6	7	9	10
user u_a Sequence	1	2	1	2	1	2	1	2	1
Port	$c_a(1)$	$c_a(2)$	$c_a(1)$	$c_a(2)$	$c_a(1)$	$c_a(2)$	$c_a(1)$	$c_a(2)$	$c_a(1)$
user u_b Sequence	-	-	1	2	3	4	1	2	3
Port	-	-	$c_b(1)$	$c_b(2)$	$c_b(3)$	$c_b(4)$	$c_b(1)$	$c_b(2)$	$c_b(3)$

Fig. 11.3 An example of OBR-2 when the users adopt a symmetric algorithm and $\delta = 2$

$$\{c_b(1), c_b(2), c_b(3), c_b(4)\} \tag{11.19}$$

When user u_b starts the process $\delta = 2$ time slots later, rendezvous can be achieved on port u_1 with $TTR = 4$ when $c_a(1) = c_b(4) = u_1$, as illustrated in Fig. 11.1.

However, it is easy to check that the above simple algorithm cannot guarantee rendezvous for all scenarios such as $\delta = 0$.

In Fig. 11.1, two users run different strategies to achieve rendezvous, which is impossible in practice since all users should run the same algorithm, i.e. symmetric algorithm. Figures 11.2 and 11.3 show another example of OBR-2 where the users share the same strategy. Similar to the above, user u_a and user u_b have the same available port sets, as in Fig. 11.1, i.e. user u_a has two available ports, and the available port set is:

$$C_a = \{c_a(1), c_a(2)\} \tag{11.20}$$

and user u_a has four and the set is:

$$C_b = \{c_b(1), c_b(2), c_b(3), c_b(4)\} \tag{11.21}$$

but only one common available port exists:

$$c_a(1) = c_b(4) = u_1 \tag{11.22}$$

Different from Fig. 11.1, both users run a same simple algorithm: each user accesses the ports by repeating the sequence:

$$\{1, 2, \ldots, k\} \tag{11.23}$$

which are of local labels, where k is the number of available ports. Thus user u_a repeats accessing the ports:

$$\{c_a(1), c_a(2), c_a(1), c_a(2), \ldots\} \tag{11.24}$$

until rendezvous, and similarly for user u_b.

For the asynchronous scenario, supposing user u_b starts the attempt $\delta = 1$ time slot later, rendezvous is achieved as depicted in Fig. 11.2 at time slot 5 since $c_a(1) = c_b(4)$. However, it is easy to see that the above simple algorithm cannot guarantee rendezvous for all scenarios such as when $\delta = 2$, as illustrated in Fig. 11.3.

Combining the two examples, we aim to design efficient distributed algorithms with bounded TTR values for different types of rendezvous settings.

11.5   Chapter Summary

In this part, we propose the oblivious blind rendezvous problem and present different types of rendezvous algorithms. Oblivious blind rendezvous assumes that the external ports are not labeled by a universal rule, and the users have to label the ports themselves locally. In Chap. 12, we design asymmetric algorithms for the users, which is similar to the blind rendezvous problem. In Chap. 13, we study symmetric algorithms for the users in a distributed system. We first assume the users are non-anonymous and they can design algorithms on the basis of the distinguishable identifiers. The method of designing fully distributed rendezvous algorithms is then presented in Chap. 14 where no global information is utilized in rendezvous, such as the number of external ports, the number of users in the system, and the maximum identifiers for the users. We study oblivious blind rendezvous for anonymous users in Chap. 15 and we introduce several randomized algorithms. Finally, we extend the oblivious blind rendezvous between two users to the rendezvous problem among multiple users in a multi-hop system in Chap. 16.

To begin with, we introduce the oblivious blind rendezvous problem, with examples. Different from blind rendezvous, the external ports are not labeled globally and the users may see different "local" labels of a pair of connected ports. In this chapter, we introduce the system model including several aspects in a rendezvous setting: $Algorithm, Time, Port$ and ID. We will present algorithms for different rendezvous settings. We also use Maximum Time to Rendezvous ($MTTR$) and Expected Time to Rendezvous ($ETTR$) to evaluate the rendezvous algorithms. These two metrics are used to evaluate the performance of the worst situation and the average performance respectively. We also provide some examples of oblivious blind rendezvous to demonstrate the differences with blind rendezvous.

References

1. Chuang, I., Wu, H.-Y., Lee, K.-R., & Kuo, Y.-H. (2013). Alternate hop-and-wait channel rendezvous method for cognitive radio networks. In *INFOCOM*.
2. ETSI. (2012). EN 301 598 white space devices (WSD); wireless access systems operating in the 470 MHz to 790 MHz frequency band.
3. Flores, R.E. Guerra, A.B., & Kightly, E.W. (2013). IEEE 802.11af: A standard for TV white space spectrum sharing. *IEEE Communications Magazine*, *62*, 92–100.
4. Ofcom. (2013). Regulatory Requirements for White Space Devices in the UHF TV band. http://www.cept.org/Documents/se-43/6161/.

Chapter 12
Asymmetric Oblivious Blind Rendezvous Algorithms

Abstract In this chapter, we present asymmetric algorithms for the oblivious blind rendezvous problem. In the setting, we fix Alg as:

$$RS =< Alg\text{-}AS, \, Time, \, Port, \, ID, \, Obli > \tag{12.1}$$

where $Time \in \{Syn, Asyn\}$, $Port \in \{Port - S, Port\text{-}AS\}$, and $ID \in \{Non\text{-}Anon, Anon\}$. Similar to designing asymmetric algorithms for the blind rendezvous problem in Chap. 6, the users' identifiers (IDs) could be used to break symmetric situations in distributed rendezvous algorithms, but they do not play the vital role, since many works do not assume the existence of distinct IDs. Therefore, we design rendezvous algorithms for 4 different rendezvous setting: *Synchronous and Port-Symmetric, Asynchronous and Port-Symmetric, Synchronous and Port-Asymmetric*, and *Asynchronous and Port-Asymmetric*, regardless of the choice of ID from $\{Non\text{-}Anon, Anon\}$. In Sect. 12.1, we present efficient algorithms for port-symmetric scenarios, where the users could start synchronously or asynchronously. In Sect. 12.2, we handle the synchronous and port-asymmetric situation, and the rendezvous algorithms for asynchronous and port-asymmetric situation are provided in Sect. 12.3. Finally, we summarize the chapter in Sect. 12.4.

12.1 Port-Symmetric Rendezvous

Consider two users u_a and u_b, and suppose their available port sets are $C_a, C_b \subseteq U$ respectively. We introduce rendezvous algorithms for port-symmetric rendezvous for both synchronous users and asynchronous users, in the following setting:

$$RS =< Alg\text{-}AS, \, Time, \, Port\text{-}S, \, ID, \, Non\text{-}Obli > \tag{12.2}$$

where $Time$ can be either Syn or $Asyn$.

© Springer Nature Singapore Pte Ltd. 2017
Z. Gu et al., *Rendezvous in Distributed Systems*,
DOI 10.1007/978-981-10-3680-4_12

Algorithm 12.1 Fixed Port Accessing Algorithm

1: Denote set of available ports as $C = \{c(1), c(2), \ldots, c(k)\}$ where $k = |C|$;
2: Choose a random number $s \in [1, k]$;
3: Access port $c(s)$ all the time;

Algorithm 12.2 Oblivious Sequential Accessing Algorithm

1: Denote time $t := 1$, the user's port set as $C \subseteq U$;
2: Denote $k := |C|$, and $C = \{c(1), c(2), \ldots, c(k)\}$
3: **while** Not rendezvous **do**
4:     Let $x := (t - 1)\%k + 1$;
5:     Access port $c(x)$ in time t;
6:     $t := t + 1$;
7: **end while**

We introduce two different algorithms. The first one is called the Fixed Port Accessing (FPA) algorithm, which is described in Algorithm 12.1. In the algorithm, the user labels the available channels as:

$$C = \{c(1), c(2), \ldots, c(k)\} \tag{12.3}$$

and it chooses a random number s in the range $[1, k]$, where k is the number of all available ports. Then, it will access port $c(s)$ all the time.

The second algorithm is called the Oblivious Sequential Accessing (OSA) algorithm which is described in Algorithm 12.2. In the algorithm, the user accesses the ports in a sequential way just like the Sequential Accessing Algorithm in Chap. 6. The difference is: the user does not know the global label of each port, and it has to access the port sequentially by its local labels.

Consider two users u_a and u_b, and suppose user u_a runs Algorithm 12.1 while user u_b runs Algorithm 12.2. We show that they can rendezvous quickly if they start at the same time.

Theorem 12.1 *Two synchronous, port-symmetric users can achieve rendezvous within k time slots, where k is the number of the available ports, if they adopt Algorithms 12.1 and 12.2 respectively.*

Proof Denote the available port sets for user u_a and u_b as:

$$\begin{cases} C_a = \{c_a(1), c_a(2), \ldots, c_a(k)\} \\ C_b = \{c_b(1), c_b(2), \ldots, c_b(k)\} \end{cases}$$

where k represents the number of available ports.

From the definition of port-symmetric in oblivious blind rendezvous (see Chap. 11), for any port $c_a(l_a) \in C_a$, there exists $l_b \in [1, k]$ for user u_b such that port $c_b(l_b)$ in the available port set C_b is connected to $c_a(l_a)$, i.e. they correspond to a common port with the same global label.

Without loss of generality, suppose user u_a adopts Algorithm 12.1 and the chosen port is always $c_a(l_a)$. Clearly, there exits $l_b \in [1, k]$ such that $c_b(l_b)$ is connected with $c_a(l_a)$. Since two users start at the same time, user u_j will access port $c_b(l_b)$ in the l_bth time slot, which causes the rendezvous to happen. Therefore, the time complexity is bounded as:

$$TTR = l_b \le k \qquad (12.4)$$

Therefore, the theorem holds.

When two users are asynchronous, one user may start its algorithm earlier. Under this situation, rendezvous can also be guaranteed in a short time. We derive the time complexity of rendezvous in the following theorem.

Theorem 12.2 *Two asynchronous, port-symmetric users can achieve rendezvous within k time slots, k is the number of available ports, if they adopt Algorithms 12.1 and 12.2 respectively.*

Proof Similar to the proof of Theorem 12.1, we suppose user u_a chooses port $c_a(l_a)$ which is connected to user u_b's port $c_b(l_b)$. We show the theorem from two aspects.

Suppose user u_a starts Δ time slots earlier than user u_b. When user u_b starts Algorithm 12.2, it will spend l_b time slots to access port $c_b(l_b)$ from its clock. Since user u_a accesses port $c_a(l_a)$ all the time, they can rendezvous in time slot l_b after user u_b starts.

Suppose user u_b starts Δ time slots earlier than user u_a. Obviously, when user u_a starts Algorithm 12.1, it will access port $c_a(l_a)$ all the time. For the first time of user u_a, user u_b is accessing port $c_b(\Delta\%k + 1)$. If $\Delta\%k + 1 \le l_b$, after $l_b - (\Delta\%k)$ time slots, user u_b will access port $c_b(l_b)$, which makes them rendezvous. If $\Delta\%k + 1 > l_b$, after $k + l_b - (\Delta\%k) \le k$ time slots, user u_b will access port $c_b(l_b)$. Therefore, they can rendezvous within k time slots if user u_b starts Δ time slots earlier.

Combining the two aspects, rendezvous between two asynchronous users can be achieved within k time slots.

For example, suppose two symmetric users have three available ports and port $c_a(2)$ is connected to port $c_b(3)$. As shown in Fig. 12.1, if they start at the same time, rendezvous happens at time slot 3; they can also achieve rendezvous in 3 time slots if one user starts later, which is shown in Fig. 12.2.

12.2 Synchronous and Port-Asymmetric Rendezvous

Consider two users u_a and u_b, and suppose their available port sets are $C_a, C_b \subseteq U$ respectively. In the following setting,

Fig. 12.1 An example of oblivious blind rendezvous between two synchronous users

Time	1	2	3	4	5	6

user u_a	Port	$c_a(2)$	$c_a(2)$	$c_a(2)$	$c_a(2)$	$c_a(2)$	$c_a(2)$

user u_b	Port	$c_b(1)$	$c_b(2)$	$c_b(3)$	$c_b(1)$	$c_b(2)$	$c_b(3)$

Fig. 12.2 An example of oblivious blind rendezvous between two asynchronous users

Time	1	2	3	4	5	6

user u_a	Port	-	-	-	$c_a(2)$	$c_a(2)$	$c_a(2)$

user u_b	Port	$c_b(1)$	$c_b(2)$	$c_b(3)$	$c_b(1)$	$c_b(2)$	$c_b(3)$

$$RS = < Alg\text{-}AS, \, Syn, \, Port\text{-}AS, \, ID, \, Obli > \tag{12.5}$$

two users start the rendezvous process at the same time and the users may have different available port sets.

Suppose user u_a labels the N external ports as:

$$\{p_a(1), p_a(2), \ldots, p_a(N)\} \tag{12.6}$$

while user u_b labels the N ports as:

$$\{p_b(1), p_b(2), \ldots, p_b(N)\} \tag{12.7}$$

We use the local labels of the users to design asymmetric algorithms.

We introduce two asymmetric algorithms. The first one is the Modified Oblivious Sequential Accessing Algorithm, which is described in Algorithm 12.3. The algorithm is a modified version of Algorithm 12.2 for port-symmetric users. The intuitive idea is also to access the ports sequentially, but the difference is: the user accesses the ports by the local labels of N channels, i.e. repeating accessing the ports by sequence:

$$\{p_a(1), p_a(2), \ldots, p_a(N)\} \tag{12.8}$$

for user u_a. Obviously, when some port $p(x)$ is not available, we also choose a random available one for replacement.

Algorithm 12.3 Modified Oblivious Sequential Accessing Algorithm

1: Denote time $t := 1$, the user's port set as $C \subseteq U$;
2: **while** Not rendezvous **do**
3: Let $x := (t - 1)\%N + 1$;
4: **if** Port $p(x)$ is available **then**
5:   Access port $p(x)$ in time t;
6: **else**
7:   Choose port $p(x)$ randomly from C;
8:   Access port $p(x)$ in time t;
9: **end if**
10: $t := t + 1$;
11: **end while**

The other algorithm is called the Oblivious Temporary Wait Algorithm, which is described in Algorithm 12.4. The algorithm is based on Algorithm 6.4 in Chap. 6. In the algorithm, the user accesses a fixed port for N time slots, and it will access the next port for the next N time slots. We illustrate the method of constructing the rendezvous sequence in Fig. 12.3. If the chosen port is not available, it will choose a random one for replacement.

Algorithm 12.4 Oblivious Temporary Wait Algorithm

1: Denote time $t := 1$, the user's port set as $C \subseteq U$;
2: **while** Not rendezvous **do**
3: Let $t' := (t - 1)\%N^2 + 1$;
4: Let $x := \lceil (t' - 1)/N \rceil + 1$;
5: **if** Port $p(x)$ does not belong to set C **then**
6:   Choose $p(x)$ randomly from C;
7: **end if**
8: Access port $p(x)$ for rendezvous attempt;
9: $t := t + 1$;
10: **end while**

Without loss of generality, suppose user u_a runs Algorithm 12.3 and user u_b runs Algorithm 12.4. We show the correctness and derive the complexity of rendezvous in the following theorem.

Fig. 12.3 The illustration of the oblivious temporary wait algorithm

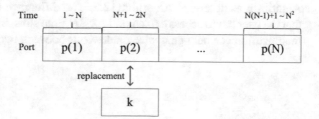

Theorem 12.3 *Two synchronous, port-asymmetric users can achieve rendezvous within N^2 time slots, if they adopt Algorithms 12.3 and 12.4 respectively.*

Proof Since user u_a and user n_b have at least one common available port, i.e. $C_a \cap C_b \neq \emptyset$, suppose port $p_a(l_a)$ is connected to $p_b(l_b)$, where $p_a(l_a) \in C_a$, $p_b(l_b) \in C_b$, $1 \leq l_a, l_b \leq N$.

Since two users start the algorithm at the same time, after $(l_b - 1) * N$ time slots, user u_b will access port $p_b(l_b)$ for the next N time slots, from Line 4 of Algorithm 12.4. It is easy to see that user u_a will access ports $\{p_a(1), p_a(2), \ldots, p_a(N)\}$ in the following N time slots. Notice that, if some port $p_a(i)$ is not available, it will replace it with a random available port. Therefore, when user u_a accesses port $p_a(l_a)$ in time slot $(l_b - 1) * N + l_a$, rendezvous happens. Therefore, rendezvous is guaranteed within $(l_b - 1) * N + l_a \leq N^2$ time slots.

12.3 Asynchronous and Port-Asymmetric Rendezvous

Consider two users u_a and u_b, and suppose their available port sets are C_a, $C_b \subseteq U$. In the following settings,

$$RS = < Alg\text{-}AS,\ Asyn,\ Port\text{-}AS,\ ID,\ Obli > \tag{12.9}$$

two users start rendezvous process in different time slots, and the users may have different available port sets.

Suppose user u_a labels the N external ports as:

$$\{p_a(1), p_a(2), \ldots, p_a(N)\} \tag{12.10}$$

while user u_b labels the N ports as:

$$\{p_b(1), p_b(2), \ldots, p_b(N)\} \tag{12.11}$$

We design a modified algorithm to achieve rendezvous on the basis of the users' local labels.

Suppose user u_a also adopts the Modified Oblivious Sequential Accessing Algorithm (Algorithm 12.3), but user u_b adopts the Modified Oblivious Temporary Wait Algorithm as described Algorithm 12.5. The difference between Algorithm 12.4 is that the user has to access a fixed port for $2N$ time slots, instead of N time slots. This modification can help guarantee rendezvous between two asynchronous users.

Algorithm 12.5 Modified Oblivious Temporary Wait Algorithm

1: Denote time $t := 1$, the user's port set as $C \subseteq U$;
2: **while** Not rendezvous **do**
3:   Let $t' := (t-1)\%2N^2 + 1$;
4:   Let $x := \lceil (t'-1)/2N \rceil + 1$;
5:   **if** Port $p(x)$ does not belong to set C **then**
6:     Choose $p(x)$ randomly from C;
7:   **end if**
8:   Access port $p(x)$ for rendezvous attempt;
9:   $t := t + 1$;
10: **end while**

Theorem 12.4 *Two asynchronous, port-asymmetric users can achieve rendezvous within $2N^2$ time slots, if they adopt Algorithms 12.3 and 12.5 respectively.*

Proof Since user u_a and user n_b have at least one common available port, i.e. $C_a \bigcap C_b \neq \emptyset$, suppose port $p_a(l_a)$ is connected to $p_b(l_b)$, where $p_a(l_a) \in C_a$, $p_b(l_b) \in C_b$, $1 \leq l_a, l_b \leq N$. We derive the theorem from two aspects.

Suppose user u_a starts earlier than user u_b. When user u_b starts the rendezvous algorithm, it will access port $p_b(l_b)$ for $2N$ time slots, after $(l_b - 1) * 2N$ time slots have elapsed. No matter by how many slots user u_a is earlier than user u_b, there must exists N time slots such that user u_a accesses ports $\{p_a(1), p_a(2), \ldots, p_a(N)\}$ sequentially (when some port is not available, we replace it by any available one), while user u_b keeps accessing port $p_b(l_b)$. Therefore, rendezvous can be guaranteed in $l_b * 2N \leq 2N^2$ time slots.

Suppose user u_a starts Δ time slots later than user u_b. Since the period of user u_b's accessing algorithm is $2N^2$, it is also $2N$ times of user u_a's repeating period. We only need to consider the situation $0 \leq Delta < 2N^2$, since the other Δ values can be thought of as repeating the rendezvous sequence.

(1) If $\lfloor \Delta/2N \rfloor < l_b$, from user u_b's clock, u_b will access port $p_b(l_b)$ from time $(p_b - 1) * 2N + 1$ to $p_b * 2N$, while user u_a will repeat accessing the external ports sequentially, which implies rendezvous must happen as discussed above.

(2) If $\lfloor \Delta/2N \rfloor > l_b$, in the next repeat when user u_b accesses port $p_b(l_b)$ for $2N$ time slots, rendezvous is also guaranteed. Therefore, the time complexity from user u_a's clock is bounded within $2N(N - (\lfloor \Delta/2N \rfloor - l_b)) \leq 2N^2$.

(3) If $\lfloor \Delta/2N \rfloor = l_b$: if $\Delta\%2N <= N$, user u_a will access N continuous ports while user u_b will access $p_b(l_b)$ during the period. This is similar as the first situation, and rendezvous happens within $2N$ time slots; if $\Delta\%2N > N$, they must rendezvous when user u_b accesses port $p_b(l_b)$ in the next repeat, which is similar to the second situation. The time is also bounded by $2N^2$ time slots.

Combining these aspects, two asynchronous users can achieve rendezvous within $2N^2$ time slots, by adopting two asymmetric algorithms.

12.4   Chapter Summary

In this chapter, we design asymmetric algorithms for the users to achieve oblivious blind rendezvous. When the users are allowed to run different algorithms, we may not use the user's identifier to break symmetry. Therefore, we mainly handle four different rendezvous settings: *Synchronous and Port-Symmetric, Asynchronous and Port-Symmetric, Synchronous and Port-Asymmetric*, and *Asynchronous and Port-Asymmetric*.

Actually, when the users are port-symmetric, we can handle both the synchronous and the asynchronous scenarios. Notice that, symmetric ports have different meanings compared with blind rendezvous. In traditional blind rendezvous, two port-symmetric users mean their available ports are the same, while two port-symmetric users in oblivious blind rendezvous mean each available port must be connected to one available port of the other user. We present two different algorithms called the Fixed Port Accessing (FPA) algorithm and the Oblivious Sequential Accessing (OSA) algorithm, where the first one keeps accessing one available port, while the other one accesses the port sequentially. Two users running two proposed algorithms respectively can achieve rendezvous within k time slots, where k is the number of available ports.

For the synchronous and port-asymmetric setting, we modify the OSA algorithm and design the oblivious version of the Temporal Wait Algorithm in Chap. 6. The intuitive idea is: one user waits for a sufficiently long time while the other accesses the ports sequentially. The method is also similar to designing asymmetric algorithms for the blind rendezvous problem. For the asynchronous and port-asymmetric settings, we also modify the Oblivious Temporary Wait algorithm and it can work well for two neighboring asynchronous and port-asymmetric users.

This chapter presents the intuitive idea of designing asymmetric algorithms for oblivious blind rendezvous, and we will introduce more symmetric algorithms for the other types of rendezvous settings.

Chapter 13
Oblivious Blind Rendezvous
for Non-anonymous Users

Abstract In this chapter, we present *symmetric algorithms* for the blind rendezvous problem between two *non-anonymous* users. In the setting, we fix Alg and ID as:

$$RS = < Alg\text{-}S, Time, Port, Non\text{-}Anon, Obli > \qquad (13.1)$$

where $Port \in \{Port - S, Port - AS\}$ and $Time \in \{Syn, Asyn\}$. It is easy to see that there are 4 different rendezvous settings when Alg is fixed as symmetric, ID is fixed as non-anonymous, and $Label$ is fixed as oblivious. The most difficult part in designing symmetric algorithms is to break symmetry among the indistinguishable users, such that the users in the system can perform differently from each other. In this chapter, we assume each user has distinct identifiers (IDs) and this information can help break symmetry in the distributed system. In Sect. 13.1, we present a rendezvous algorithm for two synchronous users when they start the oblivious rendezvous algorithm at the same time. This algorithm is not that efficient compared with the blind rendezvous algorithm for two synchronous users where the users share the same labels of the external ports. We then present two distributed algorithms for asynchronous users in Sect. 13.2. The intuitive idea is to design hopping sequences based on the differences in the users' IDs. The first algorithm (ID Hopping) utilizes the user's ID directly while the other one (Multi-Step channel Hopping) uses the converted bits of the user's ID, which is similar to the idea of designing local sequence based algorithms in Chap. 9. In order to show the efficiency of the proposed algorithms, we present the lower bounds for oblivious blind rendezvous when the users are non-anonymous in Sect. 13.3. Finally, we summarize the chapter in Sect. 13.4.

13.1 Synchronous Oblivious Blind Rendezvous

Consider two users u_a and u_b, and suppose their available port sets are $C_a, C_b \subseteq U$ respectively. For the following setting,

$$RS = < Alg\text{-}S, Syn, Port, Non\text{-}Anon, Obli > \qquad (13.2)$$

© Springer Nature Singapore Pte Ltd. 2017
Z. Gu et al., *Rendezvous in Distributed Systems*,
DOI 10.1007/978-981-10-3680-4_13

where two users have the same start time, we introduce efficient rendezvous algorithms that work well for both port-symmetric and port-asymmetric scenarios.

13.1.1 Synchronous Check and Hop Algorithm

In the first place, we introduce a simple ID conversion algorithm, Algorithm 13.1, with input identifier (ID) I and base value b. The output consists of $l + 1$ bits where each bit ranges in $[0, b)$. Algorithm 13.1 converts the user's ID to a new number under base b.

For example, input $(8, 2)$ corresponds to the output $(1, 0, 0, 0)$, which can be thought of as the common binary representation.

Algorithm 13.1 ID Conversion (I, b)

1: *Input: I, b;*
2: *Output: $d = \{d_0, d_1, \cdots, d_l\}$;*
3: $l := \lfloor log_b I \rfloor, i := l$;
4: **while** $i \geq 0$ **do**
5:     $d_i := I \mod b$;
6:     $I := \lfloor I/b \rfloor$;
7:     $i := i - 1$;
8: **end while**

We present the Synchronous Check and Hop (SCH) algorithm in Algorithm 13.2. Suppose each user has a unique ID $I \in [1, M]$, available channel set C, and an upper bound estimation of the number of total external ports $\widetilde{N} = O(N)$. Notice that, in many practical applications, the user may not know the exact number of all external ports (N of them), and the proposed algorithm also works when the user is only aware of an estimation of value N.

The SCH algorithm consists of two stages: *Synchronous Check Stage* and *Hop Stage*. The Synchronous Check Stage generates $CT = p\widetilde{P}$ numbers, as shown in Lines 9–10, where p is the smallest prime number $p \geq \max\{k, 3\}$, $k = |C|$ and $\widetilde{P}$ is the smallest prime number no less than the estimation $\widetilde{N}$. From Line 10, this stage repeats the sequence:

$$\overrightarrow{z} = \{1, 2, \ldots, p\} \tag{13.3}$$

for $\widetilde{P}$ times, which is then mapped as:

$$\overrightarrow{z'} = \{1, 2, \ldots, k, 1, 2, \ldots, p - k\} \tag{13.4}$$

as in Line 16 of the algorithm since only k ports are available. Figure 13.1 shows the process of the construction.

Algorithm 13.2 Synchronous Check & Hop Algorithm

1: *Input:* I, C, an estimation $\widetilde{N}$;
2: $k := |C|$;
3: Find the smallest prime numbers $p \geq \max\{k, 3\}$, $\widetilde{P} \geq \widetilde{N}$;
4: $l := \lfloor \log_{p-1} I \rfloor$;
5: Invoke ID Conversion $(I, p-1)$ and the output is d;
6: $D := \{d_0 + 1, d_1 + 1, \ldots, d_l + 1, 0\}$;
7: $CT := p\widetilde{P}$, $HT = p^2(l+2)$, $FL = p^2$, $t := 0$;
8: **while** Not rendezvous and $t < CT + HT$ **do**
9:   **if** $t < CT$ **then**
10:     $z = t \mod p + 1$;
11:   **else**
12:     $x = \lfloor (t - CT)/FL \rfloor$, $y = (t - CT) \mod FL$;
13:     $y_1 = \lfloor y/p \rfloor$, $y_2 = y \mod p$;
14:     $z = (y_1 + y_2 \cdot D(x)) \mod p + 1$;
15:   **end if**
16:   $z' = (z-1) \mod k + 1$, access port $c(z') \in C$;
17:   $t = t + 1$;
18: **end while**

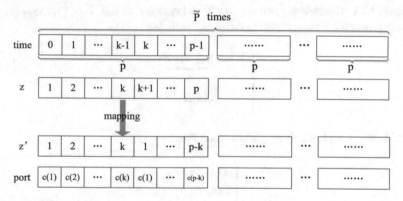

Fig. 13.1 The construction of the synchronous check stage

The Hop Stage generates $HT = p^2(l+2)$ numbers in Lines 12–14, where $l = \lfloor \log_{p-1} I \rfloor$. It consists of $l+2$ frames and the length of each frame is $FL = p^2$. In the i-th frame, the construction of first p numbers can be thought of the user hops in a circle of p nodes with labels $\{1, 2, \ldots, p\}$:

$$1 \to 1 + D(i) \to (2D(i)) \mod p + 1 \to (3(D(i)) \mod p + 1 \to \cdots \quad (13.5)$$

from Lines 13–14 of the algorithm. We call $D(i)$ the *hopping step* as the difference between two consecutive numbers. Then the next p numbers are constructed the same way as increasing the first number to 2 and holding the same hopping step $D(i)$. Thus the Hop Stage can be constructed iteratively.

For example, when $D(i) = 0$, the corresponding sequence is:

$$\vec{z} = \{\underbrace{1, 1, \ldots, 1}_{p}, \underbrace{2, 2, \ldots, 2}_{p}, \ldots, \underbrace{p, p, \ldots, p}_{p}\} \tag{13.6}$$

and when $D(i) = 1$, the corresponding sequence is:

$$\vec{z} = \{\underbrace{1, 2, \ldots, p}_{p}, \underbrace{2, 3, \ldots, p, 1}_{p}, \ldots, \underbrace{p, 1, 2, \ldots, p-1}_{p}\} \tag{13.7}$$

13.1.2   Correctness and Complexity

The intuitive idea of constructing the Synchronous Check Stage originates from the following lemma.

Lemma 13.1   *Considering two vectors* $X = \{x_1, x_2, \ldots, x_m\}$, $Y = \{y_1, y_2, \ldots, y_n\}$, *if their greatest common divisor* $gcd(m, n) = 1$, *let:*

$$\begin{cases} \hat{X} = [\underbrace{XX \cdots X}_{n}] \\ \hat{y} = [\underbrace{YY \cdots Y}_{m}] \end{cases}$$

$\forall i \in [1, m]$, $j \in [1, n]$, *there exists k such that*

$$\begin{cases} \hat{X}(k) = x_i \\ \hat{Y}(k) = y_j \end{cases}$$

Proof From the construction of $\hat{X}$ and $\hat{Y}$, when:

$$\begin{cases} k_x = i + m\theta_x, \theta_x \in [0, n) \\ k_y = j + n\theta_y, \theta_y \in [0, m) \end{cases}$$

we have:

$$\begin{cases} \hat{X}(k_x) = x_i \\ \hat{Y}(k_y) = y_j \end{cases}$$

Let $k_x = k_y$, we get:

$$i + m\theta_x = j + n\theta_y \tag{13.8}$$

Take modular operation on both sides to derive:

$$\begin{cases} i = j + n\theta_y \quad \mathrm{mod}\ m \\ i + m\theta_x = j \quad \mathrm{mod}\ n \end{cases}$$

Since $gcd(m, n) = 1$, there exist m^{-1}, n^{-1} such that:

$$\begin{cases} m \cdot m^{-1} = 1 \quad \mathrm{mod}\ n \\ n \cdot n^{-1} = 1 \quad \mathrm{mod}\ m \end{cases}$$

Therefore, we compute:

$$\begin{cases} \theta_x = (j - i) \cdot m^{-1} \quad \mathrm{mod}\ n \\ \theta_y = (i - j) \cdot n^{-1} \quad \mathrm{mod}\ m \end{cases}$$

Thus $k = k_x = k_y$ exists such that $\hat{X}(k) = x_i$ and $\hat{Y}(k) = y_j$.

From this lemma, for any two users with (I_a, C_a) and (I_b, C_b), if the corresponding prime numbers in Line 3 of Algorithm 13.2 satisfy $p_a \neq p_b$, which implies $gcd(p_a, p_b) = 1$, then rendezvous is guaranteed in the Synchronous Check Stage.

Lemma 13.2 *For two synchronous users with (I_a, C_a) and (I_b, C_b) running Algorithm 13.2, if $p_a \neq p_b$, they can achieve rendezvous in $T = \min\{CT_a, CT_b\} = \min\{p_a, p_b\}\widetilde{P}$ time slots, where $CT_a = p_a\widetilde{P}, CT_b = p_b\widetilde{P}$ as in Line 7.*

The lemma can be verified directly from Lemma 13.1. However, when $p_a = p_b$, the users may not rendezvous on a common port in the Synchronous Check Stage. Thus we design the Hop Stage to guarantee rendezvous under this specific situation.

We derive the time complexity to achieve rendezvous when the parameters $p_a = p_b$, as follows.

Lemma 13.3 *For two synchronous users with (I_a, C_a) and (I_b, C_b) running Algorithm 13.2, if $p_a = p_b = p$, rendezvous is guaranteed in $T = CT_a + \min\{HT_a, HT_b\} = p\widetilde{P} + (\min\{l_a, l_b\} + 2) \cdot p^2$ time slots, where $l_a = \lfloor \log_{p-1} I_a \rfloor$, $l_b = \lfloor \log_{p-1} I_b \rfloor$.*

Proof From Line 7 of Algorithm 13.2, we denote:

$$\begin{cases} HT_a = p_a^2(l_a + 2) \\ HT_b = p_b^2(l_b + 2) \end{cases}$$

Denote ID Conversion output of (I_a, p) and (I_b, p) in Algorithm 13.1 as:

$$\begin{cases} d_a = \{d_{a,0}, d_{a,1}, \ldots, d_{a,l_a}\} \\ d_b = \{d_{b,0}, d_{b,1}, \ldots, d_{b,l_b}\} \end{cases}$$

When two users run Algorithm 13.2, we denote the variables in Line 6 as D_a, D_b respectively.

In the first place, we show following the claim.

Claim *There exists* $\lambda \leq \min\{l_a, l_b\} + 1$ *such that* $D_a(\lambda) \neq D_b(\lambda)$.

From the construction of D_a, D_b as in Line 6,

$$D_a(l_a + 1) = D_b(l_b + 1) = 0 \tag{13.9}$$

and we have:

$$\forall i \in [0, l_a], \forall j \in [0, l_b], 0 < D_a(i) < p, 0 < D_b(j) < p \tag{13.10}$$

If $l_a \neq l_b$, without loss of generality, suppose $l_a < l_b$; let $\lambda = l_a + 1$, and then:

$$\begin{cases} D_a(\lambda) = 0 \\ D_b(\lambda) = d_{b,l_a} + 1 \geq 1 \end{cases}$$

and thus the claim is proved.

When $l_a = l_b$, we can check that there exists $0 \leq \lambda \leq l_a$ such that $d_{a,\lambda} \neq d_{b,\lambda}$.

In Algorithm 13.1, since two users must have different identifiers, i.e. $I_a \neq I_b$, thus:

$$D_a(\lambda) = d_{a,\lambda} + 1 \neq d_{b,\lambda} + 1 = D_b(\lambda) \tag{13.11}$$

Since two users have at least one common available port, i.e. $C_a \cap C_b \neq \emptyset$, for any port $u' \in C_a \cap C_b$, there exists $1 \leq i \leq k_a, 1 \leq j \leq k_b$ such that:

$$\begin{cases} c_a(i) = u' \\ c_b(j) = u' \end{cases}$$

Since two users begin the algorithm at the same time with $p_a = p_b = p$, we assume that they do not rendezvous in the first $T = CT + \lambda \cdot p^2$ time slots.

Consider the p^2 numbers in the λ-th frame of the Hop Stage. For $T < t < T + p^2$, let:

$$\begin{cases} y_1 = \lfloor (t - T)/p \rfloor \\ y_2 = (t - T) \mod p \end{cases}$$

the goal is to find time t such that:

$$\begin{cases} (y_1 + y_2 \cdot D_a(\lambda)) \mod p + 1 = i \\ (y_1 + y_2 \cdot D_b(\lambda)) \mod p + 1 = j \end{cases} \tag{13.12}$$

Combining the two equations, we derive:

$$y_2 \cdot [D_a(\lambda) - D_b(\lambda)] = i - j \quad \mod p \qquad (13.13)$$

Since $D_a(\lambda) \neq D_b(\lambda)$, the modular reverse $[D_a(\lambda) - D_b(\lambda)]^{-1}$ exists which suits:

$$[D_a(\lambda) - D_b(\lambda)] \cdot [D_a(\lambda) - D_b(\lambda)]^{-1} = 1 \quad \mod p \qquad (13.14)$$

thus we can derive:

$$y_2 = (i - j) \cdot [D_a(\lambda) - D_b(\lambda)]^{-1} \quad \mod p \qquad (13.15)$$

We plug this into Eq. (13.12) to compute y_1, and thus $t = T + y_1 p + y_2$.
   Then rendezvous is guaranteed in:

$$CT + (\lambda + 1) \cdot p^2 = p\widetilde{P} + (\min\{l_a, l_b\} + 2) \cdot p^2 \qquad (13.16)$$

time slots.
   Combining Lemmas 13.2 and 13.3, we can conclude the theorem as follows.

Theorem 13.1 *For two synchronous users with* (I_a, C_a), (I_b, C_b), *if* $C_a \bigcap C_b \neq \emptyset$, *rendezvous can be guaranteed in* $T = \min\{p_a, p_b\} \cdot \widetilde{P} + (\min\{l_a, l_b\} + 2) \cdot \min\{p_a, p_b\}^2 = O(\min\{k_a, k_b\} \cdot N)$ *time slots if* I_a, I_b *are polynomial functions of* p_a, p_b, *respectively.*

The SCH algorithm has two main advantages. First of all, the time complexity to achieve rendezvous is very small, especially when $k_a = O(1)$ or $k_b = O(1)$, the time could be merely $T = O(N)$ time slots, which is comparable to the blind rendezvous problem for synchronous users (as described in Chap. 7). Moreover, Algorithm 13.2 can terminate automatically and decide whether a common available port exists between C_a and C_b. If no rendezvous happens in either the Synchronous Check Stage or the Hop Stage, the user can claim that it does not share a common available port with the potential neighboring user.

However, there are also several disadvantages for the SCH algorithm. First, SCH cannot be applied to asynchronous users because a user does not know when the others start the algorithm, and thus it is incorrect to terminate after $CT + HT$ time slots. Second, if both users have a large number of available ports, the SCH algorithm works worse than the blind rendezvous algorithm for synchronous users. This is because the uncertainty of the external ports increases the hardness of designing efficient algorithms.

13.2   Asynchronous Oblivious Blind Rendezvous

Consider two users u_a and u_b, and suppose their available port sets are $C_a, C_b \subseteq U$ respectively. In the following setting,

$$RS = < Alg\text{-}S, Asyn, Port, Non\text{-}Anon, Obli > \qquad (13.17)$$

two users may start the rendezvous algorithm at different times. We introduce efficient rendezvous algorithms that work well for both port-symmetric and port-asymmetric scenarios.

13.2.1   ID Hopping Algorithm

In this section, we present a deterministic distributed algorithm called *ID Hopping (IDH)* for the OBR-2 problem when the users are asynchronous. Assuming each user has a distinct identifier (ID), the hopping sequence is generated on the basis of the ID and the number of available ports. Moreover, the IDH algorithm is influenced by the two global parameters: the number of all the ports N and the maximum value of the users' ID M.

13.2.1.1   Algorithm Description

Denote the user's identifier (ID) as $I \in [1, M]$ and the available port set as C. The IDH algorithm is described as in Algorithm 13.3, where a sequence of length $T = 2N\hat{P}$ is generated, and it is composed of N frames where each frame contains $2\hat{P}$ numbers, $\hat{P}$ being the smallest prime number larger than both N and M.

Algorithm 13.3 ID Hopping Algorithm

1: Find the smallest prime $\hat{P}$ such that $\hat{P} > \max\{N, M\}$;
2: $T := 2N\hat{P}, t := 0, n = |C|$;
3: **while** Not rendezvous **do**
4:     $t' := t \bmod T$;
5:     $x := \lfloor \frac{t'}{2\hat{P}} \rfloor, y := t' \bmod 2\hat{P}$;
6:     $z = (x + yI) \bmod \hat{P} + 1$;
7:     $z' = (z - 1) \bmod n + 1$, access port $c(z')$ in C;
8:     $t := t + 1$;
9: **end while**

For the i-th frame ($0 \le i < N$), the $2\hat{P}$ numbers are constructed as in Lines (5–6). We set $i + 1$ to the 0-th number and the j-th number is constructed as:

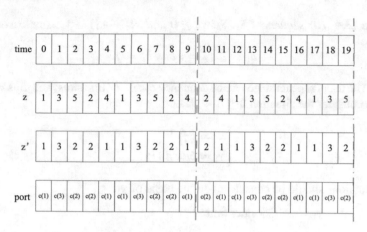

Fig. 13.2 An example of IDH (Algorithm 13.3)

$$(i + j \cdot I) \mod \hat{P} + 1 \tag{13.18}$$

This procedure can be regarded as picking numbers from a cycle with labels $\{0, 1, \ldots, \hat{P} - 1\}$, where the first number (the 0-th number) is $i + 1$ and the second one is I steps larger under the modular operation. We refer to this number as the *hopping step*. In Algorithm 13.3, the user's ID (I) is the hopping step. Since there are only n available ports, generated numbers in $[n + 1, \hat{P}]$ are mapped to $[1, n]$ to accelerate the process, which is described in Line 7.

An example of this construction is depicted in Fig. 13.2. Supposing $N = 4$, $M = 3$ and the user with identifier $I = 2$ has three available ports ($n = |C| = 3$), the sequence is constructed accordingly as shown in the figure and the ports to access in each time slot are also shown. Notice that, each frame has $2\hat{P} = 10$ time slots and we only show the first two frames in the figure.

13.2.1.2 Correctness and Complexity

For users u_a and u_b, suppose the available port sets are C_a, C_b and their IDs are I_a, I_b respectively. Denote the sequences generated in Algorithm 13.3 (as Line 6 before mapping) as

$$\begin{cases} S_a = \{a_0, a_1, \ldots, a_{T-1}\} \\ S_b = \{b_0, b_1, \ldots, b_{T-1}\} \end{cases}$$

where $T = 2N\hat{P}$. Without loss of generality, suppose user u_b is $\delta \geq 0$ time slots later than user u_a. We first derive an important lemma.

Lemma 13.4 *For sequences S_a, S_b: $\forall \delta \geq 0$ and $\forall i, j \in [1, \hat{P}]$, there exists $t < T$ such that*

$$a_{(\delta+t) \bmod T} = i \text{ and } b_t = j. \tag{13.19}$$

Proof The users repeat the generated sequence every T time slots, and thus we only need to consider the situation $0 \leq \delta < T$. Let:

$$\begin{cases} x_1 = \lfloor \frac{\delta}{2\hat{P}} \rfloor \\ y_1 = \delta \bmod 2\hat{P} \end{cases}$$

Two situations should be considered on the basis of y_1:

Case 1: $0 \leq y_1 < \hat{P}$. Consider time:

$$t = x_2 \cdot 2\hat{P} + y_2 \tag{13.20}$$

where $0 \leq x_2 < N, 0 \leq y_2 < \hat{P}$. Let:

$$x_2 + y_2 I_b + 1 \equiv j \quad \bmod \hat{P} \tag{13.21}$$

and we can compute:

$$y_2 = (j - x_2 - 1) I_b^{-1} \bmod \hat{P}. \tag{13.22}$$

Here I_b^{-1} exists such that $I_b I_b^{-1} \equiv 1 \bmod \hat{P}$ since I_b and $\hat{P}$ are co-primes. We enumerate x_2 from 0 to $N - 1$; y_2 can be computed from Eq. (13.22) correspondingly and we denote the value as y_2^h when $x_2 = h$. Then these N values comprise the set:

$$Y = \{y_2^0, y_2^1, \ldots, y_2^{N-1}\} \tag{13.23}$$

Denote the set of corresponding time slots as:

$$T_B = \{t_0, t_1, \ldots, t_{N-1}\} \tag{13.24}$$

where t_h is computed as:

$$t_h = h \cdot 2\hat{P} + y_2^h \tag{13.25}$$

It is clear that $\forall t_h \in T_B, 0 \leq h < N$, we have:

$$\begin{cases} t_h < T \\ b_{t_h} = j \end{cases}$$

Denote:

$$T_A = \{t'_0, t'_1, \ldots, t'_{N-1}\} \tag{13.26}$$

where $t'_h = (t_h + \delta) \bmod T$. We show that there exists $g \in [0, N)$ such that $a_{t'_g} = i$.

Considering any two time slots $t'_g, t'_h \in T_A$ where user u_a accesses different ports:

$$a_{t'_g} = (x_1 + g) + (y_1 + y_2^g)I_a \bmod \hat{P} + 1$$

$$a_{t'_h} = (x_1 + h) + (y_1 + y_2^h)I_a \bmod \hat{P} + 1$$

Plugging in the expression of y_2^g, y_2^h in Eq. (13.22), we can derive:

$$a_{t'_g} - a_{t'_h} \equiv (g - h)(I_a I_b^{-1} - 1) \neq 0 \bmod \hat{P}. \tag{13.27}$$

Here $I_a \neq I_b$, $I_a, I_b < \hat{P}$ implies $I_a I_b^{-1} \neq 1$. So $a_{t'_g} \neq a_{t'_h}$.

Since $|T_A| = |T_B| = N$, there are N different values for the N time slots in T_B, and thus there exists t'_g such that $a_{t'_g} = i$, which concludes the lemma.

Case 2: $\hat{P} \leq y_1 < 2\hat{P}$. Consider time:

$$t = x_2 \cdot 2\hat{P} + y_2 \tag{13.28}$$

where $0 \leq x_2 < N$ and $\hat{P} \leq b_2 < 2\hat{P}$. Using the same technique as in Case 1, we can find $t < T$ such that:

$$a_{(\delta+t) \bmod T} = i \text{ and } b_t = j \tag{13.29}$$

From the two aspects, the lemma holds.

Based on the lemma, we derive the time complexity to achieve rendezvous in Theorem 13.2.

Theorem 13.2 *Algorithm 13.3 guarantees rendezvous between two asynchronous users of the OBR-2 problem in* MTTR $= 2N\hat{P}$ *time slots, where* $\hat{P} \leq 2 \max\{N, M\}$.

Proof Since $C_a \cap C_b \neq \emptyset$, for any common available port $c^* \in C_a \cap C_b$, there exists $i \in [1, n_a]$ and $j \in [1, n_b]$ such that:

$$\begin{cases} a_i = c^* \\ b_j = c^* \end{cases}$$

where $n_a = |C_a|, n_b = |C_b|$.

Without loss of generality, supposing user u_b is δ time slots later than user u_a. From Lemma 13.4, there exists $t < T$ such that they both access port c^*, and thus rendezvous can be guaranteed in $T = 2N\hat{P}$ time slots no matter when they start the process.

Remark 13.1 P is shown to be $\hat{P} \leq 2 \max\{M, N\}$ and thus $MTTR = O(N \max \{N, M\})$. If $M = O(N)$ in Algorithm 13.3, $MTTR = O(N^2)$.

13.2.2 Multi-step Port Hopping Algorithm

The IDH algorithm works well when $M = O(N)$. However, when the number of users increases, this algorithm becomes inefficient (for example, when $M = N^3$). The reason is that the user's ID is used as the hopping step and it increases the time complexity to achieve rendezvous when M is large. Therefore, we propose a new algorithm called *Multi-Step port Hopping (MSH)* which is more efficient for systems with a large number of users. Two techniques are utilized in the algorithm: ID scaling and hopping with different steps (similar to the SCH algorithm).

13.2.2.1 Algorithm Description

Suppose the user's ID is $I \in [1, M]$ and the available port set is C, the MSH algorithm is described in Algorithms 13.4 and 13.5. First, the ID is scaled to $\lfloor \log_N M \rfloor + 1$ bits and each bit ranges from 1 to N.[1] The process is similar to the ID conversion of the SCH algorithm.

Algorithm 13.4 ID Scale Function

1: *Input: I*;
2: *Output:* $d = \{d(1), d(2), \ldots, d(l)\}$;
3: $l := \lfloor \log_N M \rfloor + 1, i := 1, cur(0) := I$;
4: **while** $i \leq l$ **do**
5:    $d(i) := cur(i - 1) \bmod N + 1$;
6:    $cur(i) := \lfloor cur(i - 1)/N \rfloor$
7:    $i := i + 1$;
8: **end while**

For example, for $N = 8$, $M = 100$, $I = 30$, the scaled values are:

$$d = \{7, 4, 1\} \tag{13.30}$$

The scale function plays a key role in the rendezvous algorithm design and the scaled values are used as the hopping steps in Algorithm 13.5.

---

[1] Here, each 'bit' does not mean 0 or 1, but a value in $[1, N]$.

Algorithm 13.5 Multi-Step Port Hopping Algorithm

1: Find the smallest prime P such that $P > N$;
2: $T := 2NP, t := 0, n = |C|, l := \lfloor \log_N M \rfloor + 1$;
3: Invoke Algorithm 13.4 on the user's ID and get the output $d = \{d(1), d(2), \ldots, d(l)\}$;
4: **while** Not rendezvous **do**
5:   **if** $t < T$ **then**
6:     $z := \lfloor t/2P \rfloor + 1$;
7:   **else**
8:     $t' := (t - T) \bmod (2lT)$;
9:     $x := \lfloor t'/2T \rfloor + 1, y := t' \bmod 2T$;
10:     $y_1 := y \bmod (2P), y_2 := (\lfloor y/(2P) \rfloor) \bmod N + 1$;
11:     $z := (y_2 + y_1 \cdot d(x) - 1) \bmod P + 1$;
12:   **end if**
13:   $z' := (z - 1) \bmod n + 1$, access port $c(z')$ in C;
14:   $t := t + 1$;
15: **end while**

Algorithm 13.5 can be thought of as generating two types of sequences. The first one is called Scale Sequence (SS) which is composed of 0 and repetitions of l scaled values. Since two users can start the rendezvous process asynchronously, bit 0 is added as a special flag to represent the start of the user. We represent this type of sequence as:

$$SS = \{0, \underbrace{d(1), d(2) \ldots, d(l)}_{l}, \underbrace{d(1), d(2), \ldots, d(l)}_{l}, \ldots \ldots \} \tag{13.31}$$

The other one is called Port Hopping Sequence which is composed of different frames based on SS, as shown in Fig. 13.3. Actually, there are $N + 1$ different types of frames:

$$F(0), F(1), \ldots, F(N) \tag{13.32}$$

and each type is composed of N segments.

For example, $F(i)$ has N segments and each segment contains $2P$ numbers. The 0-th number of the j-th segment is constructed as j and the k-th number is:

$$(j + k * i - 1) \bmod P + 1 \tag{13.33}$$

We can find that, the construction of each segment of $F(i)$ can be seen as accessing a port in $[1, P]$ by hopping i steps.

For example, $F(0)$ and $F(1)$ are constructed as follows:

$$F(0) = \underbrace{1, 1, \ldots, 1}_{2P}, \underbrace{2, 2, \ldots, 2}_{2P}, \ldots, \underbrace{N, N, \ldots, N}_{2P} \tag{13.34}$$

$$F(1) = \underbrace{1, 2, \ldots, P}_{2P}, \underbrace{2, 3, \ldots, P, 1}_{2P}, \ldots, \underbrace{N, N + 1, \ldots, N - 1}_{2P} \tag{13.35}$$

Fig. 13.3 Construction of
port hopping sequence

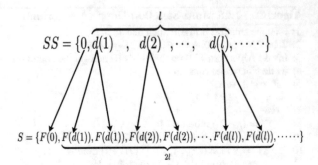

As shown in Fig. 13.3, the first number 0 is a special symbol because it does not
appear in other positions of SS and it corresponds to $F(0)$ only once, while the other
numbers in SS correspond to each type of frames twice.

13.2.2.2   Correctness and Complexity

Supposing users u_a and u_b run the MSH algorithm (Algorithm 13.5) with their
local information (C_a, I_a) and (C_b, I_b) where $C_a \cap C_b \neq \emptyset$, $I_a \neq I_b$, and let $n_a = |C_a|$, $n_b = |C_b|$, we denote:

$$\begin{cases} d_a = \{d_a(1), d_a(2), \ldots, d_a(l)\} \\ d_b = \{d_b(1), d_b(2), \ldots, d_b(l)\} \end{cases}$$

as the outputs of ID Scale function, denote SS_a, SS_b as the scale sequences that are
constructed as in the above, and denote:

$$\begin{cases} S_a = \{a_0, a_1, \ldots, a_t, \ldots\} \\ S_b = \{b_0, b_1, \ldots, b_t, \ldots\} \end{cases}$$

as the Port Hopping Sequences.

Without loss of generality, suppose user u_b starts the process $\delta \geq 0$ time slots later
than user u_a. we show the following Lemmas 13.6, 13.7 and 13.8.

Lemma 13.5 *There exists $1 \leq i \leq l$ such that $d_a(i) \neq d_b(i)$.*

The lemma can be derived easily since the IDs for two users are different.

Lemma 13.6 *Consider SS_a, SS_b: $\forall \delta' \in Z$, there exists $i \geq 0, i + \delta' \geq 0$ such that:*

$$SS_a(i) \neq SS_b(i + \delta) \tag{13.36}$$

Proof There are three cases according to different δ values:

Fig. 13.4 Example of Lemma 13.7. The block labeled *gray* represents the intersection part between two users

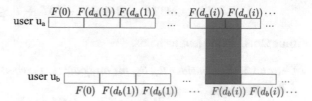

Case 1: $\delta > 0$. Let $i = 0$, $SS_a(0) = 0$ but $SS_b(\delta) \neq 0$. From Line 5 of Algorithm 13.4, each bit of $SS_b(i)$ ranges from 1 to N when $i > 0$, and thus $SS_a(0) \neq SS_b(\delta)$.

Case 2: $\delta < 0$. Let $i = -\delta > 0$, similar to the first case, we have: $SS_a(-\delta) \neq SS_b(0)$.

Case 3: $\delta = 0$. Since $SS_a(0) = SS_b(0) = 0$, the first two cases do not work here. In Lines 5,6 of Algorithm 13.4, these variables are kept as:

$$\{d_a(i), d_b(i), cur_a(i), cur_b(i)\} \tag{13.37}$$

respectively. Without loss of generality, suppose $I_a > I_b$. Find the smallest value j such that $cur_b(j) = 0$. $\forall 1 \leq i \leq j$, the conditions $d_a(i) = d_b(i)$ and $cur_a = cur_b$ cannot happen, otherwise we can conclude $I_a = I_b$ from the construction of $cur(i), d(i)$ values. Thus, if there exists $1 \leq i \leq j$ such that $d_a(i) \neq d_b(i)$, the lemma holds; if $\forall 1 \leq i \leq j$, $d_a(i) = d_b(i)$, $cur_a(j) \neq cur_b(j)$ implies $d_a(j + 1) \neq d_b(j + 1)$.

Combining the three situations, the lemma holds.

Lemma 13.7 *Consider S_a, S_b; for any pair (i, j) where $1 \leq i \leq n_a$, $1 \leq j \leq n_b$, if $0 \leq \delta < T$,*

$$\exists t \leq 2lT \text{ s.t. } a_{(\delta+t)} = i \text{ and } b_t = j. \tag{13.38}$$

Proof From Lemma 13.5, there exists $1 \leq i \leq l$ such that

$$d_a(i) \neq d_b(i) \tag{13.39}$$

From the construction of S_a, S_b as shown in Fig. 13.3, we show that rendezvous is guaranteed in the $(2i)$-th frame (i.e., the first $F(d_b(i))$) of S_b, as depicted in Fig. 13.4.

Since $\delta < T$, the first frame $F(0)$ of sequence S_b intersects with frame $F(0)$ of S_a, and thus the $2i$-th frame (the first $F(d_b(i))$ in Fig. 13.4) intersects with two $F(d_a(i))$ frames in S_a. From the construction of each $F(d_a(i))$ and each $F(d_b(i))$, their hopping steps are $d_a(i)$, $d_b(i)$ respectively, and thus it is similar to the sequences of Algorithm 13.3 where the input IDs are $d_a(i) \neq d_b(i)$ and $\delta < T$.

Therefore we can conclude that: from Theorem 13.2, rendezvous can be guaranteed in $2NP = T$ time slots if we consider the start time of user u_b to be in the beginning of $F(d_b(i))$; then rendezvous is guaranteed in:

$$MTTR \le (1 + 2(i - 1))T + T \le 2lT \tag{13.40}$$

time slots. So the lemma holds.

Lemma 13.8 *Consider* S_a, S_b, *for any pair* (i, j) *where* $1 \le i \le n_a$, $1 \le j \le n_b$, *if* $\delta \ge T$,

$$\exists t \le T \ s.t. \ a_{(\delta + t)} = i \ and \ b_t = j. \tag{13.41}$$

Proof Suppose the constructed Scale Sequence for user u_a is:

$$SS_a = \{0, d_a(1), d_a(2), \ldots, d_a(l), \ldots\} \tag{13.42}$$

Let:

$$x = \lfloor \delta/T \rfloor + 1 \tag{13.43}$$

$$y = (\delta - T) \bmod T \tag{13.44}$$

There are four cases to be analyzed.

Case 1: $0 \le y < T - 2jP$. We show that there exists such time $t \le T$ and $a_{(\delta+t)}$ belongs to x-th Frame of S_a.
From the construction of frame $F(0)$ in S_b, $\forall t \in [2(j - 1)P, 2jP)$, we have:

$$b_t = j \tag{13.45}$$

Let:

$$t' = \delta + t \tag{13.46}$$

$$\lfloor t'/T \rfloor + 1 = x \tag{13.47}$$

the $2P$ numbers $a_{t'}$ belong to the x-th frame of sequence S_a. From the construction of each frame except $F(0)$, every $2P$ consecutive numbers contain every number in $[1, P]$ at least once, and thus we can find time $t \in [2(j - 1)P, 2jP)$ such that:

$$\begin{cases} b_t = j \\ a_{(\delta+t)} = i \end{cases}$$

Case 2: $T - 2jP \le y < T - (2j - 1)P$. Consider last P numbers of the x-th frame of S_a: it is clear that each number in $[1, P]$ appears once including i. When we consider:

$$xT - P \le \delta + t < xT \tag{13.48}$$

$t \in [2(j-1)P, 2jP)$ implies $b_t = j$, and thus there exists such $t <$ $2jP \leq T$ satisfying the lemma.

Case 3: $T - (2j-1)P \leq y < T - 2(j-1)P$. Consider the first P numbers of the $(x+1)$-th frame of S_a; similar to Case 2, the corresponding P numbers in S_b are always j, and thus such $t < T$ exists.

Case 4: $y \geq T - 2(j-1)P. \forall t \in [2(j-1)P, 2jP), b_t = j$ and the $2P$ numbers $a_{(\delta+t)}$ belong to the $(x+1)$-th frame of S_a. Similar to Case 1, such t can be found with $a_{(\delta+t)} = i$.

Combining the four cases, the lemma can be concluded.

We derive the time complexity of the algorithm in the following theorem.

Theorem 13.3 *The MSH algorithm (Algorithm 13.5) guarantees oblivious rendezvous for two asynchronous users in* $MTTR = 4lNP = O(N^2 \log_N M)$ *time slots, where* $P \leq 2N$.

Proof As assumed, $G = C_a \cap C_b \neq \emptyset$, and suppose $c^* \in G$. There exists $1 \leq i \leq n_a$, $1 \leq j \leq n_b$ such that:

$$\begin{cases} c_a(i) = c^* \\ c_b(j) = c^* \end{cases}$$

Without loss of generality, suppose user u_b starts the process δ time slots later. If $\delta < T$, from Lemma 13.7, rendezvous is guaranteed in $2lT$ time slots; if $\delta \geq T$, rendezvous is guaranteed in T time slots, and thus we derive the maximum time to rendezvous which is bounded by:

$$MTTR \leq 2lT = 4lNP = O(N^2 \log_N M) \tag{13.49}$$

time slots. Then, the theorem holds.

Generally speaking, if value M is bounded by a polynomial function of the total number of external ports N, the length of scaled bits is a constant and two users can be guaranteed to rendezvous in $O(N^2)$ time slots. Moreover, this result is also comparable to even state-of-the-art non-oblivious rendezvous algorithms, as shown in Table 8.1.

Remark 13.2 When $M = O(N)$, Algorithm 13.3 works better than Algorithm 13.5. However, when the number of users increases substantially, where M can be much larger than N, Algorithm 13.5 then works much better than Algorithm 13.3. In general, if $M = poly(N)$ and Algorithm 13.5 can guarantee rendezvous within $MTTR = O(N^2)$ time slots.

13.3 Lower Bound for Oblivious Blind Rendezvous

As presented in Sects. 13.1 and 13.2, distributed algorithms can be designed for oblivious blind rendezvous between two users. In this chapter, we derive the lower bound on any oblivious rendezvous algorithm for two non-anonymous users to show the efficiency of the proposed algorithms. In this section, we introduce the Adversary Assignment Graph (AAG)[1] where an adversary can assign the universal ports to any local labeled port freely. We will use the AAG to derive the lower bounds on any rendezvous algorithm for the oblivious blind rendezvous problem.

13.3.1   Adversary Assignment Graph

Since the users do not know the label of each port in the universal port set:

$$U = \{u_1, u_2, \ldots, u_N\} \tag{13.50}$$

they have to label the available ports locally and two (even non-anonymous) users cannot know the other's labels. Therefore, we assume an adversary exists in the system who can assign the universal port to any port with local labels for the users. Rendezvous is achieved in the worst scenario when two users can rendezvous on some common available port for every port assignment by the adversary.

Suppose two users (u_a and u_b) have distinct identifiers $I_a, I_b (I_a \neq I_b)$ and available port sets C_a, C_b respectively. Denote $k_a = |C_a|, k_b = |C_b|$, and denote:

$$\begin{cases} C_a = \{c_a(1), c_a(2), \ldots c_a(k_a)\} \\ C_b = \{c_b(1), c_b(2), \ldots, c_b(k_b)\} \end{cases}$$

Assume both users run the same deterministic algorithm $\mathscr{F}$ and denote a_t, b_t as the outputs of both users at time slot t. Then we introduce the construction of the *Adversary Assignment Graph (AAG)* as follows: (for simplicity, suppose both users start $\mathscr{F}$ at the same time):

(1) Initially, there are two rows of nodes in the graph and the upper row contains k_A separated nodes while the lower row contains k_B separated nodes. The nodes in the upper row represent the available ports of user u_a, while the nodes in the lower row represent the available ports of user u_b;

(2) for each time slot t, after the users run the rendezvous algorithm $\mathscr{F}$ with outputs $a_t \in [1, k_a], b_t \in [1, k_b]$, connect the a_t-th node of the upper row to the b_t-th node of the lower row;

(3) the adversary can assign any universal ports to C_a, C_b satisfying:

$$\begin{cases} \forall 1 \leq i, j \leq k_a, c_a(i) \neq c_a(j) \\ \forall 1 \leq i, j \leq k_b, c_b(i) \neq c_b(j) \end{cases}$$

Fig. 13.5 An example of adversary assignment graph: rendezvous is not achieved

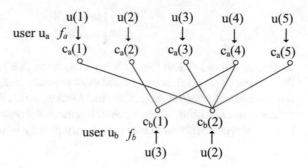

Fig. 13.6 An example of adversary assignment graph: rendezvous is achieved

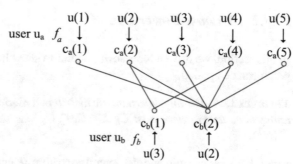

(4) if for every assignment in the third step, there exists a common universal port that is connected in the graph, rendezvous is achieved, otherwise continue the process from the second step.

For example, given five universal ports:

$$U = \{u_1, u_2, u_3, u_4, u_5\} \tag{13.51}$$

and user u_a and u_b have 5 and 2 available ports respectively.

As depicted in Fig. 13.5, there are two rows of nodes in the graph and the upper row has 5 nodes while the lower row has only 2. For the first time slot, if the outputs of the rendezvous algorithm for both users are:

$$\begin{cases} a_0 = 1 \\ b_0 = 2 \end{cases}$$

connect the first node in the upper row to the second node in the lower node as in the figure. Suppose at some time slot t, there are 6 pairs of nodes connected as depicted in the figure. We say rendezvous is achieved if for every assignment by the adversary, at least one common available port is connected. However, Fig. 13.5 shows the port assignment in which no common available port is connected, which implies rendezvous does not happen under this situation.

Suppose another pair of node is connected where:

$$\begin{cases} a_{t+1} = 2 \\ b_{t+1} = 2 \end{cases}$$

as seen as the red line in Fig. 13.6; we can check that for every port assignment by the adversary, at least one common port is connected, which implies rendezvous must happen. Therefore, we can use the AAG to derive the lower bound on oblivious rendezvous algorithms for two non-anonymous users, by finding the smallest t such that rendezvous happens in every adversary port assignment.

13.3.2   A Loose Lower Bound

In this section, we drive a loose lower bound to show how the adversary assignment graph can be applied.

Theorem 13.4 *For any deterministic distributed algorithm solving OBR-2 for non-anonymous users, there exist $C_a, C_b, C_a \cap C_b \neq \emptyset$ such that the MTTR value is $\Omega(N^2)$.*

Proof For any deterministic distributed algorithm $\mathscr{F}$ and based on the set of available port set C and the identifier I, we have:

$$f \mapsto [1, n] \tag{13.52}$$

where $n = |C|$

Suppose users u_a and u_b have different IDs $I_a \neq I_b$. We let:

$$|C_a| = |C_b| = \lceil N/2 \rceil, |C_a \cap C_b| = 1 \tag{13.53}$$

Equivalently, denote the only common port between the users as c^*, and there exists $1 \leq i, j \leq \lceil N/2 \rceil$ such that:

$$c_a(i) = c_b(j) = c^* \tag{13.54}$$

We construct the Adversary Assignment Graph (AAG) as in Fig. 13.7 where two rows of nodes exist in the graph and the number of nodes in each row is exactly $n = \lceil N/2 \rceil$. The upper row represents user u_a's local labels of the available ports with indices $\{1, 2, \ldots, n\}$ and the lower row represents user u_b's labels.

Let a_t, b_t be the outputs of the algorithm for users u_a and u_b at time slot t respectively; we know that:

$$a_t = f(a_1, a_2, \ldots, a_{t-1}, n, I_a)$$
$$b_t = f(b_1, b_2, \ldots, b_{t-1}, n, I_b)$$

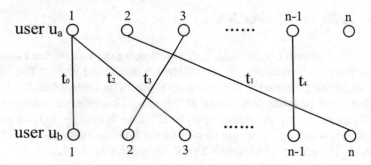

Fig. 13.7 Adversary assignment graph for Theorem 13.4

Without loss of generality, suppose user u_b begins δ slots later; accordingly, we connect node $a_{t+\delta}$ in the upper row with b_t in the lower row with an edge having the label t. Notice that, if the two nodes are already connected, then we just update the label on the edge.

For example, $(1, 1)$ is connected in t_0 as depicted in Fig. 13.7 and $(2, n)$, $(1, 3)$, $(3, 2)$, $(n - 1, n - 1)$ are also connected.

Suppose there exists an adversary who can assign global ports from the set:

$$U = \{u_1, u_2, \ldots, u_N\} \tag{13.55}$$

to C_a and C_b; rendezvous will not be achieved if the common port c^* in the upper row is not connected to c^* in the lower row. Since the inputs to the algorithm $\mathscr{F}$ are fixed, where the inputs for user u_a are I_a and $|C_a|$, and the inputs for user u_b are I_b and $|C_b|$. Clearly, the lower bound on the maximum time to rendezvous ($MTTR$) is the smallest value T such that (c^*, c^*) is connected in every adversary assignment.

Let δ_a be the smallest degree among all the upper nodes. Obviously, if $\delta_a < n$, the adversary can find a node i in the upper row and j in the lower row such that (i, j) is not connected, and then assigns c^* to both of them, which implies rendezvous is not achieved. After the assignment of port c^*, it is easy to assign the other non-intersecting ports to other nodes. We can verify that $\delta_a < n$ exists if $T < n^2$, which can be deduced easily. Thus, C_a and C_b can be constructed by the adversary such that rendezvous does not happen. Therefore, the lower bound on the maximum time to rendezvous can be deduced as:

$$MTTR = n^2 = \Omega(N^2) \tag{13.56}$$

Then, the theorem holds.

13.3.3 A Refined Lower Bound

The bound in Theorem 13.4 is loose. In this section, we derive a refined lower bound based on the users' available port sets. Since the users have distinct IDs, different sequences may be generated by different users to access the ports, which implies good algorithms could be made with a small *MTTR* value. However, we demonstrate that the *MTTR* value for any algorithms could be $\Omega((k_a - k_g) \cdot (k_b - k_g))$ for the worst case situation, where k_a, k_b are the number of available ports for two asynchronous users and k_g is the number of common available ports they share.

Theorem 13.5 *For any deterministic algorithm solving the oblivious rendezvous problem between two users with (I_a, C_a), (I_b, C_b), $C_a \bigcap C_b \neq \emptyset$, $I_a \neq I_b$, there exist mappings:*

$$f_a : C_a \mapsto U_a \subseteq U$$
$$f_b : C_b \mapsto U_b \subseteq U$$

such that the maximum time to rendezvous is:

$$\Omega((k_a - k_g) \cdot (k_b - k_g)) \tag{13.57}$$

where $k_a = |C_a|, k_b = |C_b|, k_g = |C_a \bigcap C_b|$.

Proof For any deterministic algorithm $\mathscr{F}$, let a_t, b_t be the ports to access in time slot t when the users run the algorithm with inputs (I_a, C_a), (I_b, C_b), respectively. It's obvious that:

$$a_t = \mathscr{F}(a_0, a_1, \dots, a_{t-1}, I_a, C_a)$$
$$b_t = \mathscr{F}(b_0, b_1, \dots, b_{t-1}, I_b, C_b)$$

Consider the scenario when two users (u_a and u_b) start $\mathscr{F}$ at the same time; without loss of generality, let:

$$U_g = \{u_1, u_2, \dots, u_g\} \tag{13.58}$$

be the common available ports they share.

We also use the Adversary Assignment Graph (AAG) to derive the lower bound. There are two rows of nodes in the graph. The number of the nodes in the upper row is k_a while the other row's number of nodes is k_b; each row represents the available ports of each user. For each time slot t, connect (a_t, b_t) in the graph if they are not connected, where a_t corresponds to the node in the upper row and b_t is in the lower row. As shown in Fig. 13.8,

$$(c_a(1), c_b(1)), (c_a(1), c_b(3)), (c_a(2), c_b(2)), (c_a(2), c_b(3)), (c_a(3), c_b(2)), \dots \tag{13.59}$$

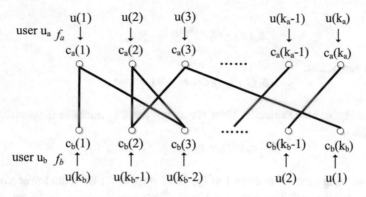

Fig. 13.8 Adversary assignment graph for Theorem 13.5

are connected and at most one edge is added to the graph in each time slot.

Assume there exists an adversary who can assign any port $c_a(i) \in C_a$ or $c_b(j) \in C_b$ to any port with global label $u' \in U$ at any time slot t. As shown in Fig. 13.8, the adversary maps every port in C_a as:

$$f_a : c_a(i) \mapsto u(i) \tag{13.60}$$

and maps every port in C_b as:

$$f_b : c_b(i) \mapsto u(k_b + 1 - i) \tag{13.61}$$

Rendezvous is not achieved if there exists an assignment such that $\forall u' \in U_g$, u' in the upper row is not connected to u' in the lower row. Thus the lower bound of the *MTTR* value is the smallest t such that for every adversary assignment, there exists $u' \in U_g$ where (u', u') is connected in the graph.

We demonstrate that rendezvous cannot be guaranteed in

$$t < (k_a - k_g)(k_b - k_g) \tag{13.62}$$

time slots. Denote sets:

$$A_t = \{a_1, a_2, \ldots, a_t\}$$
$$B_t = \{b_1, b_2, \ldots, b_t\}$$

and we construct the AAG as described above. Let $\delta_a(i)$ be the degree of node $c_a(i)$ and we sort these nodes of the upper row in ascending order:

$$c_a(1'), c_a(2'), \ldots, c_a(k_a') \tag{13.63}$$

where

$$\delta_a(1') \le \delta_a(2') \le \cdots \le \delta_a(k_a') \tag{13.64}$$

We can verify that:

$$\delta_a(i') \le (k_b - k_g), \forall 1 \le i \le k_g \tag{13.65}$$

from the Pigeonhole Principle. Then, we assign these k_g nodes to the ports in U_g as:

$$f_a : c_a(i') \mapsto u_i \in U_g, \forall 1 \le i \le k_g. \tag{13.66}$$

Considering i increases from 1 to k_g, find a node $c_b(\hat{i})$ from the lower row corresponding to node $c_a(i')$ such that $(c_a(i'), c_b(\hat{i}))$ is not connected in the graph. Since $\delta_a(i') \le k_b - k_g$, there are at least k_g nodes not connected to node $c_a(i')$. However, at most $i - 1 < k_g$ nodes of the lower row are assigned, and thus such a node exists. We assign this node as:

$$f_b : c_b(\hat{i}) \mapsto u_i \in U_g, \forall 1 \le i \le k_g. \tag{13.67}$$

Finally, assign all other nodes to $U \setminus U_g$ as:

$$f_a : c_a(i') \mapsto u' \in U_a', \forall k_g < i \le k_a \tag{13.68}$$

$$f_a : c_b(\hat{i}) \mapsto u' \in U_b', \forall k_g < i \le k_b \tag{13.69}$$

where $c_b(\hat{i})$ represents the nodes that have not been assigned and $U_a', U_b' \subseteq U \setminus U_g, U_a' \cap U_b' = \emptyset$. Thus, such an adversary assignment exists, and it insinuates that rendezvous is not achieved. Hence, we conclude that the *MTTR* value for any deterministic algorithm is $\Omega((k_a - k_g) \cdot (k_b - k_g))$, which concludes the theorem.

Remark 13.3 The IDH and MSH algorithms are comparable with the derived lower bound in Theorem 13.4 when $M = O(N)$ and M is bounded by a polynomial function of N.

13.4 Chapter Summary

In this chapter, we study the oblivious blind rendezvous problem between two non-anonymous users where the users are assumed to have been assigned unique identifiers (IDs). Different from the previous chapter, we design symmetric algorithms for non-anonymous users, where they have to run the same rendezvous algorithm. We introduce rendezvous algorithms for synchronous users and for asynchronous users.

We first handle oblivious blind rendezvous for two synchronous, non-anonymous users. Assuming each user has a distinguishable ID, we design the Synchronous Check and Hop (SCH) algorithm, which consists of two stages: the synchronous check stage repeats accessing ports sequentially for a sufficiently long time, and the hop stage generates rendezvous hopping sequence on the basis of the converted bits of the user's ID. The synchronous check stage can guarantee rendezvous between two users if the numbers of two users' available ports are much different. The hop stage is designed for two users whose numbers of available ports are about the same. Therefore, the algorithm is suitable for two synchronous users no matter how many available ports they have.

However, the synchronous check stage of the SCH algorithm is inapplicable for two asynchronous users. Therefore, we handle the oblivious blind rendezvous problem for two asynchronous users by designing rendezvous sequence on the basis of the users' ID. We propose two asynchronous rendezvous algorithms. The first one is called the ID Hopping (IDH) algorithm, which utilizes the user's ID to design hopping sequence. Therefore, the algorithm is related to both N (the number of all ports) and M (the maximum ID value), which guarantees rendezvous within $O(N \max\{N, M\})$ time slots.

In order to decrease the time complexity to achieve rendezvous, we introduce another algorithm called the Multi-Step port Hopping (MSH), which guarantees rendezvous within in $O(N^2 \log_N M)$ time slots. The idea is to convert the user's ID to a string of bits and any two distinguishable IDs correspond to two different strings of bits. Generally, when M is a polynomial function of N, the MSH algorithm can guarantee rendezvous within $O(N^2)$ time slots, which is a good result compared with non-oblivious rendezvous algorithms.

In order to show the efficiency of our described algorithm, we show the method of deriving lower bounds on oblivious blind rendezvous between two non-oblivious users. We introduce the adversary assignment graph, where an adversary is assumed to assign local labels to the available ports. By assigning different local labels, we derive the maximum time such that any adversary assignment will finally lead to rendezvous. A loose lower bound shows that $\Omega(N^2)$ time slots are needed, and a refined lower bound requires $\Omega((k_a - k_g)(k_b - k_g))$ time slots, where k_a, k_b represent two users' available ports and k_g means their common available ports. From these lower bounds, the proposed algorithms can work efficiently and guarantee rendezvous between two non-anonymous users in a reasonable time.

Reference

1. Gu, Z., Hua, Q.-S., & Dai, W. (2014). Fully distributed algorithms for blind rendezvous in cognitive radio networks. In *MOBIHOC*.

Chapter 14
Fully Distributed Rendezvous Algorithm for Non-anonymous Users

Abstract In Chap. 13, we present efficient distributed algorithms for both synchronous and asynchronous users that are non-anonymous. These algorithms utilize global information such as the number of the external ports N and the number of users M (or the maximum value for the users' identifier (ID)). In practical large scale networks, it is difficult for the users to know these information beforehand. For example, in cognitive radio networks, no general standard exists dividing the total licensed spectrum into N channels, such as the IEEE 802.11 standard which only concerns frequencies ranging 470–710 MHz [1], and so it is impractical for the users to know the value of N. Moreover, all users are physically dispersed in the system and they may join or leave freely, and hence they cannot know the number of users in advance as there is no central controller. Therefore, it is desirable to design a *fully distributed algorithm* where only the users' local information would be utilized. Actually, in a general distributed system, this kind of local information is limited to the user's ID and the number of the user's available ports since there exists no global labels for the ports. In Sect. 14.1, we present the first fully distributed algorithm called the Conversion Based Hopping (CBH) algorithm, which guarantees oblivious blind rendezvous in a short time. The correctness and complexity are analyzed in Sect. 14.2. We summarize the chapter in Sect. 14.3.

14.1 Conversion Based Hopping Algorithm

The SCH algorithm in the preceding chapter cannot work for two asynchronous users because the synchronous check stage could not work when the users start at different time slots. However, we can use the intuitive idea of the hop stage of the SCH algorithm to design distributed algorithms for two asynchronous users. Moreover, the SCH algorithm assumes each user has an estimation of N but the proposed algorithm in this chapter (we called *Conversion Based Hopping Algorithm*, or CBH for short) only uses the user's local information: the ID and the number of available ports.

Suppose the user's ID is I and the available port set is C. The CBH algorithm is described in Algorithm 14.1. With local input (I, C), Algorithm 14.1 finds the smallest prime number $p \geq \max\{k, 3\}$ where $k = |C|$ and invokes ID Conversion $(I, p - 1)$ to get the results d. The ID Conversion is described in Algorithm 13.1.

© Springer Nature Singapore Pte Ltd. 2017
Z. Gu et al., *Rendezvous in Distributed Systems*,
DOI 10.1007/978-981-10-3680-4_14

Algorithm 14.1 Conversion Based Hopping Algorithm

1: *Input: I, C*;
2: $k := |C|$;
3: Find the smallest prime numbers $p \geq \max\{k, 3\}$;
4: $l := \lfloor \log_{p-1} I \rfloor$;
5: Invoke ID Conversion $(I, p - 1)$ and the output is d;
6: **if** $(l + 2) \mod 2 = 0$ **then**
7: $l_p := l + 2; D := \{0, d_0 + 1, d_1 + 1, \ldots, d_l + 1\}$
8: **else**
9: $l_p := l + 3, D := \{0, 1, d_0 + 1, d_1 + 1, \ldots, d_l + 1\}$
10: **end if**
11: $T := 2l_p \cdot p^2, FL := 2l_p \cdot p, SL = 2p$;
12: **while** Not rendezvous **do**
13: $t' := t \mod T$;
14: $x := \lfloor t'/FL \rfloor, x' = t' \mod FL$;
15: $y_1 := \lfloor x'/SL \rfloor, y_2 = x' \mod SL$;
16: $z := x + D(y_1) \cdot y_2 \mod p + 1$;
17: $z' := (z - 1) \mod k + 1$, access port $c(z') \in C$;
18: $t = t + 1$;
19: **end while**

Then we construct the array D containing l_p numbers as in Lines 6–10, where l_p is defined to be an even number, which is different from Algorithm 13.2. Following the preprocessing, Algorithm 14.1 generates a sequence of length $T = 2l_p \cdot p^2$ as in Lines 13–16. This sequence consists of p frames of equal length $FL = 2l_p \cdot p$, where each frame contains l_p segments of length $SL = 2p$. In Line 17 of the algorithm, the sequence is mapped from $[1, p]$ to $[1, k]$ and the corresponding port is accessed by the user.

We illustrate the construction of the sequence in Fig. 14.1. It consists of p frames:

$$\{F_0, F_1, \ldots, F_{p-1}\} \tag{14.1}$$

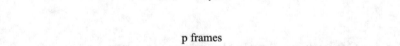

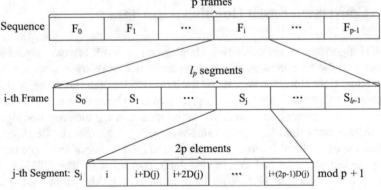

Fig. 14.1 The construction of the $T = 2l_p \cdot p^2$ sequence in CBH (Algorithm 14.1)

and each frame has l_p segments:

$$\{S_0, S_1, \ldots, S_{l_p-1}\} \tag{14.2}$$

The way to generate segment S_j of frame F_i is to construct $2p$ numbers, starting with i and the hopping step is $D(j)$; then the k-th number is constructed as such:

$$(i + kD(j)) \mod p + 1 \tag{14.3}$$

Each segment contains $2p$ numbers and this is to eliminate the asynchronous situation through doubling the length, which is similar to the method of transforming time slots into slot-aligned scenario.

There are two intuitive ideas in designing the CBH algorithm. The first one comes from the SCH algorithm when the corresponding prime numbers of the two users in Line 3 satisfy $p_a \neq p_b$, and each user repeating its own ports can guarantee rendezvous. When $p_a = p_b$, distinct IDs have different representations through the ID conversion, thus accessing the ports with these hopping steps may assure rendezvous. The proposed CBH algorithm combines these two principles and it has good performance as analyzed in the next section.

14.2 Correctness and Complexity

Assume two asynchronous users (u_a and u_b) run Algorithm 14.1 with inputs (I_a, C_a) and (I_b, C_b) where $C_a \cap C_b \neq \emptyset$, $I_a \neq I_b$ ($I_a, I_b \in [1, M]$). Without loss of generality, suppose user u_b is $\delta \geq 0$ time slots later. Denote the variables used for two users in Algorithm 14.1 as:

$$\begin{cases} (k_a, p_a, l_a, l_{p_a}, D_a, T_a, FL_a, SL_a, t_a) \\ (k_b, p_b, l_b, l_{p_b}, D_b, T_b, FL_b, SL_b, t_b) \end{cases}$$

Since $C_a \cap C_b \neq \emptyset$, there exists a port with global label $u' \in C_a \cap C_b$ and there exist $1 \leq i \leq k_a$, $1 \leq j \leq k_b$ such that

$$\begin{cases} c_a(i) = u' \\ c_b(j) = u' \end{cases}$$

We derive the time complexity to achieve rendezvous based on the following three situations:

(1) $p_a = p_b = p$ and $l_{p_a} = l_{p_b} = l_p$;
(2) $p_a = p_b = p$ but $l_{p_a} \neq l_{p_b}$;
(3) $p_a \neq p_b$;

Lemma 14.1 *If $p_a = p_b = p$ and $l_{p_a} = l_{p_b} = l_p$, rendezvous between users u_a and u_b can be guaranteed in $T = 2l_p \cdot p^2$ time slots.*

Proof If $0 \leq \delta \mod 2p < p$, there exists $x^* \geq 0, 0 \leq y_1^* < l_p, 0 \leq y_2^* < p$ such that:

$$\delta = x^* \cdot (2pl_p) + y_1^* \cdot (2p) + y_2^* \tag{14.4}$$

Suppose users u_a and u_b can achieve rendezvous on port u' at time t_a, t_b respectively, and there exists $x(a), x(b) > 0, 0 \leq y_1(a), y_1(b) < l_p, 0 \leq y_2(a) < 2p, 0 \leq y_2(b) < p$ such that:

$$t_a = x(a) \cdot (2pl_p) + y_1(a) \cdot (2p) + y_2(a) \tag{14.5}$$
$$t_b = x(b) \cdot (2pl_p) + y_1(b) \cdot (2p) + y_2(b) \tag{14.6}$$

From Lines 13–16 of Algorithm 14.1, the corresponding z values for two users could be generated to be i, j, thus:

$$x(a) + D_a(y_1(a)) \cdot y_2(a) \mod p + 1 = i \tag{14.7}$$
$$x(b) + D_b(y_1(b)) \cdot y_2(b) \mod p + 1 = j \tag{14.8}$$

Since user u_b is δ time slots later, we rewrite it as:

$$t_a = t_b + \delta \tag{14.9}$$

Plug Eqs. (14.4)–(14.6), we can get:

$$[x(a) - x(b) - x^*] \cdot (2pl) + [y_1(a) - y_1(b) - y_1^*] \cdot (2p)$$
$$+ [y_2(a) - y_2(b) - y_2^*] = 0 \tag{14.10}$$

Since $y_2(b) \in [0, p)$, $y_2(a) - y_2(b) - y_2^* = 0$. Combining this with Eqs. (14.7)–(14.8), we can derive:

$$[D_a(y_1(a)) - D_b(y_1(b))] \cdot y_2(b) + D_a(y_1(a)) \cdot y_2^* =$$
$$i - x(a) - (j - x(b)) \mod p \tag{14.11}$$

If we can find values $y_1(a), y_1(b)$ satisfying:

$$\begin{cases} D(y_1(a)) - D(y_1(b)) \neq 0 \\ y_1(a) - y_1(b) - y_1^* \mod l_p = 0 \end{cases}$$

Equation (14.11) can be solved under the constraint Eq. (14.10). We compute $y_1(a), y_1(b)$ as follows:

$$\begin{cases} y_1(a) = y_1(b) = k & \text{If } y_1^* = 0 \\ y_1(a) = y_1^*, y_1(b) = 0 & \text{If } 0 < y_1^* \le l_p - 1 \end{cases} \tag{14.12}$$

If $y_1^* = 0$, there exist $1 \le k \le l_p - 1$ such that $D_a(k) \ne D_b(k)$ from ID conversion. If $0 < y_1^* \le l_p - 1$, $D_a(y_1(a)) - D_b(y_1(b)) = D_a(y_1^*) > 0$. Thus such $y_1(a), y_1(b)$ exist and $y_1(a) - y_1(b) - y_1^* = 0$.

Since $D_a(y_1(a)) - D_b(y_1(b)) \ne 0$, $y_2(b)$ can be computed from Eq. (14.11) as follows. We plug in equation:

$$x(a) - x(b) = x^* \tag{14.13}$$

from the constraint Eq. (14.10). Then, we compute:

$$x(b) = j - 1 - D_b(y_1(b)) \cdot y_2(b) \mod p \tag{14.14}$$

and thus $x(b) \in [0, p)$. So the time to rendezvous is:

$$TTR = t_b = x(b) \cdot (2pl_p) + y_1(b) \cdot (2p) + y_2(b) \tag{14.15}$$

and it is bounded by $2l_p \cdot p^2$.

For example, users u_a and u_b have inputs $I_a = 5$, $|C_a| = 4$, $I_b = 20$, $|C_b| = 5$ and $c_a(2) = c_b(4)$ is their only common available port. Thus $p_a = p_b = 5$, $l_a = l_b = 4$ and $D_a = \{0, 1, 2, 2\}$, $D_b = \{0, 2, 2, 0\}$.

Let $\delta = 2014$ and it can be rewritten as:

$$\delta = 50 \cdot 40 + 1 \cdot 10 + 4 \tag{14.16}$$

Thus we compute the values according to Eq. (14.4) as:

$$x^* = 50, y_1^* = 1, y_2^* = 4 \tag{14.17}$$

Since $y_1^* = 1$, from Eq. (14.12), we know:

$$y_1(a) = y_1^* = 1$$
$$y_1(b) = 0$$
$$x(a) - x(b) = x^* = 50$$

From Eq. (14.11), $y_2(b) = 4$ and $x(b) = 3$. Thus

$$t_b = 3 * 40 + 4 = 124$$
$$t_a = t_b + \delta = 2138$$

We can check that user u_a accesses port $c_a(2)$ and u_b accesses port $c_b(4)$ at the same time.

If $p \leq \delta \mod 2p < 2p$, the *TTR* value is also bounded by $2l_p \cdot p^2$ time slots using the same technique above. We omit the details and the readers may deduce this situation. Therefore, the lemma holds.

Lemma 14.2 *If $p_a = p_b = p$ but $l_{p_a} \neq l_{p_b}$, rendezvous between users u_a and u_b can be guaranteed in $T = 2 \min\{l_{p_a}, l_{p_b}\} \cdot p^2$ time slots.*

Proof If $0 \leq \delta \mod 2p < p$, there exists $x^* \geq 0, 0 \leq y_1^* < l_{p_a}, 0 \leq y_2^* < p$ such that:

$$\delta = x^* \cdot (2pl_{p_a}) + y_1^* \cdot (2p) + y_2^* \tag{14.18}$$

Suppose two users can rendezvous on port u' at time t_a, t_b respectively, we have:

$$t_a = x(a) \cdot (2pl_{p_a}) + y_1(a) \cdot (2p) + y_2(a)$$
$$t_b = x(b) \cdot (2pl_{p_b}) + y_2(b) \cdot (2p) + y_2(b)$$

where $x(a), x(b) > 0, 0 \leq y_1(a) < l_{p_a}, 0 \leq y_1(b) < l_{p_b}, 0 \leq y_2(a) < 2p, 0 \leq y_2(b) < p$. Combining these with $t_a = t_b + \delta$ to derive:

$$[l_{p_a}x(a) - l_{p_b}x(b) - l_{p_a}x^* + y_1(a) - y_1(b) - y_1^*] \cdot 2p$$
$$+ y_2(a) - y_2(b) - y_2^* = 0$$

Similarly, we have:

$$\begin{cases} l_{p_a} \cdot x(a) - l_{p_b} \cdot x(b) - l_{p_a} \cdot x^* + y_1(a) - y_1(b) - y_1^* = 0 \\ y_2(a) - y_2(b) - y_2^* = 0 \end{cases} \tag{14.19}$$

We can also formulate Eqs. (14.7)–(14.8). If $l_{p_a} > l_{p_b}$, we let:

$$\begin{cases} y_1(a) = 0 \\ y_1(b) = k \neq 0 \end{cases}$$

From Eq. (14.7), we can derive:

$$x(a) = (i - 1) \mod p \tag{14.20}$$

Plugging this into Eq. (14.19), we have:

$$l_{p_b}x(b) = [l_{p_a}x(a) - l_{p_a}x^* - y_1(b) - y_1^*] \mod p \tag{14.21}$$

Since l_{p_b} is an even number, $x(b) \in [0, p)$ can be computed obviously. Then from Eq. (14.8), $y_2(b) \in [0, p)$ can be derived. Therefore, the TTR value is:

$$t_b = x(b) \cdot (2pl_{p_b}) + y_1(b) \cdot (2p) + y_2(b) \leq 2p^2 l_{p_b} \tag{14.22}$$

If $l_{p_a} < l_{p_b}$, we can bound the time to rendezvous as:

$$TTR = t_a - \delta = (x(a) - x^*) \cdot (2pl_{p_a}) + (y_1(a) - y_1^*) \cdot (2p) + (y_2(a) - y_2^*) \leq 2p^2 l_{p_a} \tag{14.23}$$

Thus, $MTTR \leq 2\min\{l_{p_a}, l_{p_b}\}p^2$.

If $p \leq \delta \mod 2p < 2p$, the TTR value is also bounded by:

$$T = 2\min\{l_{p_a}, l_{p_b}\} \cdot p^2 \tag{14.24}$$

time slots using the same technique above. Thus the lemma holds.

Lemma 14.3 *If $p_a \neq p_b$, rendezvous between users u_a and u_b can be guaranteed in $T = 2l_p \cdot p^2$ time slots, where $p = \max\{p_a, p_b\}$ and l_p is the corresponding value from $\{l_{p_a}, l_{p_b}\}$.*

Proof This lemma can be concluded similarly. Suppose $p_a < p_b$, we can derive the following equations:

$$x(a) + D_a(y_1(a)) \cdot y_2(a) \mod p_a + 1 = i$$
$$x(b) + D_b(y_1(b)) \cdot y_2(b) \mod p_b + 1 = j$$

Let $y_1(b) = 0$, then:

$$x(b) = (j - 1) \mod p_b \tag{14.25}$$

and $y_2(b) \in [0, 2p_b)$. Suppose:

$$x(a) = i' \mod p_a \tag{14.26}$$

and $y_1(a) \neq 0$, then $y_2(a)$ exists. Since $t_a = t_b + \delta$ and we know:

$$t_a = x(a) \cdot (2p_a l_{p_a}) + y_1(a) \cdot (2p_a) + y_2(a)$$
$$t_b = x(b) \cdot (2p_b l_{p_b}) + y_2(b) \cdot (2p_b) + y_2(b)$$

We can find value:

$$x(a) = i' + v(a)p_a \tag{14.27}$$

satisfying $\delta_b(v_b) + \delta - \delta_a(v_a) \in [2p_a - 2p_b, T_a)$ where $T_a = 2p_a^2 l_{p_a}$ is define as above, where:

$$\delta_b(v_b) = (2p_b l_{p_b}) \cdot (j - 1 + v(b)p_b)$$
$$\delta_a(v_a) = (2p_a l_{p_a}) \cdot x(a)$$

Obviously, we can compute:

$$\delta_b(0) \mod T_a \geq 2p_a - 2p_b \tag{14.28}$$

We let:

$$v(b) = 0$$
$$v(a) = \lfloor (\delta_b(0) + \delta)/T_a \rfloor$$
$$i' = \lfloor (\delta_b(0) + \delta - v(a)T_a)/FL_a \rfloor$$

where $FL = 2p_a l_{p_a}$; then variables $y_1(a)$, $y_2(a)$, $y_2(b)$ can be determined. Thus, the time to rendezvous can be computed as:

$$TTR = t_b = x(b) \cdot (2p_b l_{p_b}) + y_1(b) \cdot (2p_b) + y_2(b) \le 2p_b^2 l_{p_b} \tag{14.29}$$

If $p_a > p_b$, we can derive the time complexity using a similar technique:

$$TTR \le 2p_a^2 l_{p_a} \tag{14.30}$$

Therefore, the lemma holds.

Combining Lemmas 14.1–14.3, we can conclude:

Theorem 14.1 *Two users running the CBH algorithm (Algorithm 14.1) can achieve rendezvous in* $MTTR = 2l_p \cdot p^2$ *time slots where* $p = \max\{p_a, p_b\}$ *and* l_p *is the corresponding value from* $\{l_{p_a}, l_{p_b}\}$.

This theorem reveals that: the CBH algorithm can guarantee oblivious blind rendezvous between two users in a short time and it is comparable to the lower bound in Theorem 13.5 for most cases. More precisely, $MTTR = 2l_p \cdot p^2 = O(k^2)$ time slots if l_p is a constant, which implies the corresponding ID is a polynomial function of p, where $k = \max\{k_a, k_b\}$.

For example, if $k_a > k_b$ (which implies $p_a \ge p_b$), and I_a is bounded by $I_a \le p_a^c$ where c can be an arbitrary large constant, $MTTR = 2l_{p_a} \cdot p_a^2 = O(k_a^2)$. If $k_b = \Theta(k_a)$ and $k_g = o(k_a)$, the $MTTR$ value is comparable with the lower bound in Theorem 13.5 (see Chap. 13.3).

Remark 14.1 In Lemma 14.3, the $MTTR$ value can be bounded by $2l_p p^2$ time slots where $p = \min\{p_a, p_b\}$ for most cases: when $p_a < p_b$, if

$$(2p_a l_a j) \mod T_b \ge 2(p_b - p_a) \tag{14.31}$$

or

$$T_a \mod T_b = \Omega(p_b) \tag{14.32}$$

the $MTTR$ value could be very small.

14.3 Chapter Summary

In this chapter, we present the first fully distributed rendezvous algorithm for two non-anonymous users, called Conversion Based Hopping (CBH). The CBH algorithm only utilizes the user's local information: the ID and the number of available ports and it is independent of the global parameters: the number of all ports N and the maximum value for the users' ID M.

The CBH algorithm combines the intuitive idea of the MLS algorithm (in Chap. 9) and the SCH/MSH algorithm (in Chap. 13): the user's ID is first scaled (converted) to a new number given the base value that is related to the number of available ports; then the algorithm constructs hopping sequences by different hopping steps. The CBH algorithm guarantees rendezvous in $O((\max\{|C_a|, |C_b|\})^2)$ time slots under most circumstances where C_a, C_b represent the sets of two users' available ports. When the number of available ports is small, the CBH algorithm outperforms some state-of-the-art global sequence based rendezvous algorithms.

The CBH algorithm has many advantages when compared with traditional non-oblivious blind rendezvous algorithms:

(1) The CBH algorithm uses very little information. Only the user's ID and the number of available ports are used in designing the CBH algorithm. It does not require global information, such as the number of ports, the maximum ID value, the labels of the ports. Some traditional non-oblivious blind rendezvous algorithm may not utilize the user's ID either, but they may need the value of all ports, or the labels of these ports.
(2) The CBH algorithm is also suitable for non-oblivious setting, where the external ports have global labels. Compared with the state-of-the-art rendezvous algorithms, the CBH algorithm has good performance when the users' number of available ports is small.

Reference

1. A. B. Flores, R. E. Guerra, and E. W. Kightly. IEEE 802.11af: A Standard for TV White Space Spectrum Sharing. *IEEE Communications Magazine*, 2013.

Chapter 15
Oblivious Blind Rendezvous for Anonymous Users

Abstract In this chapter, we present *symmetric algorithms* for the blind rendezvous problem between two *anonymous* users. In the setting, we fix *Alg* and *ID* as:

$$RS = < Alg - S, Time, Port, Anon, Obli > \qquad (15.1)$$

where $Port \in \{Port - S, Port - AS\}$ and $Time \in \{Syn, Asyn\}$. It is easy to see that there are 4 different rendezvous settings when *Alg* is fixed as symmetric, *ID* is fixed as anonymous, and *Label* is fixed as oblivious. Different from Chaps. 13 and 14, we assume the users have no distinct identifiers to break symmetry in distributed computing. This anonymous setting makes the oblivious blind rendezvous problem difficult. In Sect. 15.1, we show the hardness due to such anonymity which gives rise to the result that no deterministic algorithm could exist for the oblivious blind rendezvous problem. Then, we present in Sect. 15.2 an efficient randomized algorithm for two *port-symmetric* users no matter whether they are synchronous or asynchronous, which achieves short expected time to rendezvous. For the most difficult setting, where the users are *port-asymmetric*, we present randomized algorithms that work well for both synchronous and asynchronous users. Finally, we summarize the chapter in Sect. 15.4.

15.1 Hardness of Anonymity

In this section, we show that there is no deterministic distributed algorithm for the OBR problem between two anonymous users, i.e. the users do not have unique identifiers (IDs) or distinguishable information.

Theorem 15.1 *There is no deterministic distributed algorithm for the OBR problem between two anonymous users.*

Proof Suppose there exists such a deterministic algorithm:

$$F : f \mapsto [1, N] \qquad (15.2)$$

for the OBR problem between two anonymous users. Consider two users A and B with two available port sets C_A, C_B satisfying

$$k_a = k_b = \lceil N/2 \rceil + 1 \tag{15.3}$$

where $k_a = |C_A|, k_b = |C_B|$ represent the number of available ports for each user; we set:

$$\forall i, c_a(i) \neq c_b(i) \tag{15.4}$$

where $c_a(i)$ and $c_b(i)$ represent the ports that are labeled as i locally by users A and B respectively.

Since $k_a + k_b > N$, at least one common available port exists between sets C_A and C_B, which is the elementary condition that the users could rendezvous. However, we show that the algorithm F cannot guarantee rendezvous.

Let $\delta = 0$, i.e. two users start the rendezvous algorithm F at the same time, and denote a_t, b_t as the labels of the ports to access in time slot t respectively; thus:

$$a_t = f(a_1, a_2, \ldots, a_{t-1}, k_a, N)$$
$$b_t = f(b_1, b_2, \ldots, b_{t-1}, k_b, N)$$

since the chosen port in time slot t is only related to the user's local information: the local labels of the ports, the number of available ports, the number of all ports N (notice that, N may not be known in advance, but we show the theorem even when this value is available for the users).

We prove that users A and B will choose the ports with the same local label $a_t = b_t$ in time slot t through the inductive method.

(1) When both users start the rendezvous algorithm, they only find out that there are $\lceil N/2 \rceil + 1$ available ports and they are indistinguishable from each other. Thus they will make the same choice and access the port with the same local label in the first time slot, i.e. $a_0 = b_0$;

(2) Suppose $a_i = b_i$ when $0 \leq i \leq t - 1$. Then user A should access port a_t as:

$$a_t = f(a_1, a_2, \ldots, a_{t-1}, k_a, N) \tag{15.5}$$

while user B should access port b_t with label:

$$b_t = f(b_1, b_2, \ldots, b_{t-1}, k_b, N) \tag{15.6}$$

Since $a_i = b_i$ when $0 \leq i \leq t - 1$, $k_a = k_b$, both users have the same input to the deterministic algorithm F and the outputs of the algorithm should be the same, i.e. $a_t = b_t$.

Combining the two aspects, both users A and B choose the ports with the same local label $a_t = b_t$ for any time slot t. Since:

$$c_a(i) \neq c_b(i), \forall i \in [1, \lceil N/2 \rceil + 1] \tag{15.7}$$

(obviously, this setting can be easy fulfilled). Rendezvous never happens even for two synchronous users. Therefore, no deterministic algorithm exists for the OBR problem between two anonymous users.

15.2 Port-Symmetric Rendezvous

In this section, we handle the port-symmetric rendezvous where two users have the same set of available ports. For simplicity, we assume all ports are available and it is easy to extend the algorithm to the general port-symmetric setting.

Recall the telephone coordination problem [1] (introduced in Sect. 1): two users A and B are isolated in two rooms and there are N telephones in each of them. The telephones are pairwise connected in some unknown fashion. For simplicity, assuming the telephones are labeled $\{1, 2, \ldots N\}$ randomly (locally) for each user, and telephone $i \in [1, N]$ of user A is connected to a certain telephone j of user B, but they do not know the connection pattern. Time is also assumed to be divided into slots of equal length and the user can select one telephone in each time slot by sending a "hello" message. If they pick a pair of connected telephones in the same time slot, they can hear from each other and it is called *rendezvous* (all time slots are regarded as aligned). Time to rendezvous (*TTR*) denotes the time cost when all users have begun the selection process and the objective is to minimize the expected time to rendezvous (*ETTR*).

When all the ports are available for the two anonymous users, it is similar to the telephone coordination problem. One simple and intuitive idea is random selection, where each user selects a random port to attempt rendezvous. This method has expected time to rendezvous (*ETTR*) as N time slots and it seems to be the best solution.

However, a better algorithm called the Anderson-Weber strategy (AW) is proposed in [2]; for two *synchronous* users and it works as follows:

(1) Choose a random value $i \in [1, N]$ and pick the i-th telephone in the first time slot;
(2) choose a constant $p \in [0, 1]$ and the user picks the i-th telephone for the next $N - 1$ time slot with probability p, or picks the telephones in the next $N - 1$ time slots according to a random permutation of set $\{1, 2, \ldots, i - 1, i + 1, \ldots, N\}$ (with probability $1 - p$);
(3) if rendezvous does not happen, repeat the second step.

It has been proved that the AW strategy is optimal when $N = 2, p = \frac{1}{2}$ (this is also shown in [3]) and $N = 3, p = \frac{1}{3}$ [1, 4, 5, 7]. It has also been conjectured that AW is asymptotically optimal when $N \geq 4$ (specifically, $ETTR = 0.8289N$ and $p = 0.2475$ when $N \to \infty$). In [6], it is proved that the AW strategy is not optimal when $N = 4$ and to find an optimal algorithm even for two synchronous users is still an open problem. In addition, the AW strategy does not work for asynchronous users. In this section, we present a randomized algorithm which works well for both synchronous and asynchronous users.

Before we describe the algorithm, we present some useful results from probability theory.

Let A be an event, $Pr(A)$ denote the probability event A happens and $Pr(\overline{A}) = 1 - Pr(A)$ the probability that event A does not happen. Let $\{B_1, B_2, \ldots, B_n\}$ be a set of disjoint events whose union is the entire sample space; then according to the law of total probability:

$$Pr(A) = \sum_{i=1}^{n} Pr(A \cap B_i) = \sum_{i=1}^{n} Pr(A|B_i) \cdot Pr(B_i) \tag{15.8}$$

Suppose X is a random variable and denote $E(X)$ as the expectation of X. If events $\{B_1, B_2, \ldots, B_n\}$ are mutually exclusive and exhaustive, according to the law of total expectation:

$$E(X) = \sum_{i=1}^{n} E(X|B_i) \cdot Pr(B_i) \tag{15.9}$$

Let $[N]$ denote the set $\{1, 2, \ldots, N\}$, and A_N^k be the number of methods selecting k elements out of $[N]$:

$$A_N^k = N(N-1)\cdots(N-k+1) \tag{15.10}$$

15.2.1 Intuitive Ideas

To begin, we show a lower bound of the expected time to rendezvous (*ETTR*) when two users are allowed to use asymmetric strategies (i.e. different algorithms). Then we derive $ETTR = N$ for the random selection algorithm. Combining the two results, we then describe the intuitive ideas in designing the proposed randomized distributed algorithm.

Lemma 15.1 *For any distributed algorithm solving the OBR problem between two anonymous users, the expected time to rendezvous satisfies:*

$$ETTR \geq \frac{N+1}{2} \tag{15.11}$$

even when the users are allowed to use asymmetric algorithms.

Proof This lemma can be derived as in [2]. Let r_t be the event that two users select the same universal port (the local labels of the port may be different) in the t-th time slot. Without loss of generality, suppose user A starts later than user B; t is the time stamp of user A since time to rendezvous (*TTR*) records the time cost when two users have both begun the process.

Since the users do not know the other's labels of the ports, we have:

$$Pr(r_t) = \frac{1}{N} \tag{15.12}$$

Note that, r_t means they can rendezvous in the t-th time slot, but not necessarily for the first time. Thus the probability two users rendezvous in the first t time slots can be bounded as:

$$Pr(r_1 \bigcup r_2 \bigcup \cdots \bigcup r_t) \leq \min \left\{ 1, \sum_{i=1}^{t} Pr(r_i) \right\} = \min\{1, \frac{t}{N}\} \tag{15.13}$$

The bound on the right side of the inequality is achieved by the strategy $\mathscr{S}$:

* One user accesses a fixed port all the time, while the other user hops through the ports according to a random permutation of $[N]$.

Obviously, we can derive the expected time to rendezvous for this strategy as:

$$ETTR = \frac{\sum_{i=1}^{n} i}{N} = \frac{N+1}{2} \tag{15.14}$$

and thus the lemma holds.

Although the strategy $\mathscr{S}$ can guarantee fast rendezvous for two anonymous users, it is inapplicable to the OBR-2 problem since two anonymous users cannot decide which role to take.

When it comes to the situation in which two asynchronous users should run a symmetric algorithm, random selection seems to be reasonable, which can be denoted as $\mathscr{R}$:

* Each user accesses a port randomly in each time slot.

We derive the expected time to rendezvous and show the efficiency of the strategy.

Lemma 15.2 *$\mathscr{R}$ has expected time to rendezvous ETTR $= N$ for two asynchronous users.*

Proof Let r_t be the event that the users access the same universal port (the local labels may be different) in the t-th time slot. Since both users access the port randomly, we can derive:

$$Pr(r_t) = \frac{1}{N} \tag{15.15}$$

Let r'_t be the event that the users can rendezvous in the t-th time slot for the first time; then we have:

$$Pr(r'_t) = Pr(\overline{r_1} \bigcap \overline{r_2} \bigcap \cdots \bigcap \overline{r_{t-1}} \bigcap r_t) = \left(1 - \frac{1}{N}\right)^{(t-1)} \cdot \frac{1}{N} \qquad (15.16)$$

Therefore, we can compute the *ETTR* value as:

$$ETTR = \sum_{t=1}^{\infty} t \cdot Pr(r'_t) = \sum_{t=1}^{\infty} t \cdot \left(1 - \frac{1}{N}\right)^{(t-1)} \cdot \frac{1}{N} = N \qquad (15.17)$$

So the lemma holds.

Lemma 15.3 *$\mathscr{R}$ guarantees rendezvous in $O(N \log N)$ time slots for two asynchronous users with high probability.*

Proof As shown in Lemma 15.2, the probability to rendezvous in each time slot t is $Pr(r_t) = \frac{1}{N}$. Since strategy $\mathscr{R}$ accesses the ports randomly for every time slot, events r_t, r'_t are independent for any $t \neq t'$. So the probability that they do not rendezvous in $cN \log N$ (c is a constant) time slots is bounded by:

$$Pr(\overline{r_1} \bigcap \overline{r_2} \bigcap \cdots \bigcap \overline{r_{cN \log N}}) = \left(1 - \frac{1}{N}\right)^{cN \log N} \qquad (15.18)$$

When $N \to \infty$, we derive that:

$$Pr(\overline{r_1} \bigcap \overline{r_2} \bigcap \cdots \bigcap \overline{r_{cN \log N}}) = e^{-c \log N} = \frac{1}{N^c} \qquad (15.19)$$

Therefore, rendezvous happens in $O(N \log N)$ time slots with high probability $1 - \frac{1}{N^c}$, which concludes the lemma.

Though strategy $\mathscr{S}$ designs asymmetric algorithms for two users, the idea that one user waits while the other user hops through all ports provides an important foundation for designing efficient randomized algorithms. The strategy $\mathscr{R}$ seems to be the best randomized algorithm where we use pure randomization in making decisions. However, if we could combine both intuitions to design randomized algorithms, we may achieve better results.

15.2.2   Stay or Random Selection Algorithm

In the section, we introduce a simple randomized distributed algorithm called *Stay or Random Selection (SRS)* that achieves rendezvous faster than random selection ($\mathscr{R}$).

As shown in Algorithm 15.1, the user makes a choice at the beginning of each *block*, which is defined as N consecutive time slots. If the chosen random value $p' \leq p$ (p is a constant we need to compute and define), the user accesses a random port and waits at it for a block of time slots; otherwise a random permutation of $[N]$ is generated and the user accesses the corresponding port in the permutation for each time slot of the block. We denote the first choice as the *stay pattern* and the second one as the *jump pattern*. The user keeps this process until rendezvous.

Algorithm 15.1 Stay or Random Selection Algorithm

1: p is a pre-defined constant in $[0, 1]$;
2: **while** Not rendezvous **do**
3:   Select a random value $p' \in [0, 1]$;
4:   **if** $p' \leq p$ **then**
5:     Select a random number in $[N]$ and access the corresponding port for the following N time slots;
6:   **else**
7:     Generate a random permutation of $[N]$ and access the corresponding ports in the following N time slots according to the permutation;
8:   **end if**
9: **end while**

The intuitive ideas of $\mathscr{S}$ and $\mathscr{R}$ are combined in our algorithm. Although the description of the algorithm is simple, finding the optimal value of p that minimizes the *ETTR* is very difficult. Compared with the AW strategy for the telephone coordination problem, our algorithm also works for two asynchronous users, which is not treated in existing works.

15.2.3 Synchronous Users Scenario

The SRS algorithm is applicable for both synchronous and asynchronous users. In this section, we analyze the rendezvous efficiency for two synchronous users and compute the appropriate p value in Algorithm 15.1.

In the synchronous situation, two users start the algorithm at the same time. As shown in Algorithm 15.1, time is divided into blocks of length N. At the beginning of each block, the user decides to be in the stay or jump pattern. Denote $r(S, J)$ as the event that user A is in the stay pattern and user B is in the jump pattern. The other three events are denoted as $r(S, S), r(J, S), r(J, J)$ similarly. Denote the expected time to rendezvous (*ETTR*) for synchronous users as T_s which can be formulated as:

$$T_s = E(S, J)Pr(S, J) + E(J, S)Pr(J, S) + E(S, S)Pr(S, S) + E(J, J)Pr(J, J) \quad (15.20)$$

where $Pr(S, J)$ is the probability that event $r(S, J)$ happens, $Pr(J, S)$ the probability that event $r(J, S)$ happens, $Pr(S, S)$ the probability that event $r(S, S)$ happens and $Pr(J, J)$ the probability that event $r(J, J)$ happens. Similarly, $E(S, J)$ is the expected time to rendezvous if user A is in the stay pattern and user B is in the jump pattern; $E(J, S)$ is the expected time to rendezvous if user A is in the jump pattern and user

B in the stay pattern; $E(S, S)$ is the expected time to rendezvous if user A is in the stay pattern and user B is in the stay pattern; and $E(J, J)$ is the expected time to rendezvous if user A is in the jump pattern and user B is in the jump pattern.

We first analyze the *ETTR* values for the four events respectively.

(1) Event $r(S, S)$:

When both users choose the stay pattern, the only chance to rendezvous is that the ports they select represent the same global port. Thus the probability to rendezvous is:

$$Pr(S, S) = \frac{1}{N} \qquad (15.21)$$

and 1 time slot is needed when rendezvous happens. Therefore

$$E(S, S) = \frac{1}{N} \cdot 1 + \left(1 - \frac{1}{N}\right)(N + T_s) \qquad (15.22)$$

(2) Event $r(S, J)$ and $r(J, S)$:

When one user chooses the stay pattern while the other one is in the jump pattern, rendezvous happens for certain:

$$Pr(S, J) = Pr(J, S) = 1 \qquad (15.23)$$

and the expected time to rendezvous is:

$$E(S, J) = E(J, S) = \frac{N + 1}{2} \qquad (15.24)$$

(3) Event $r(J, J)$:

When two users are both in the jump pattern, the expected rendezvous time is formulated as in Lemma 15.4.

Lemma 15.4 *If two users are both in the jump pattern, the expected rendezvous time is:*

$$E(J, J) = (N + 1)(1 - p(N + 1, 0) - p(N, 0)) + p(N, 0)(N + T_s) \qquad (15.25)$$

where $p(N, 0)$ is the probability that the users cannot rendezvous according to two random permutations of $[N]$ they generate respectively.

Proof Let J_1, J_2 be the permutations of $[N]$ that the users generate respectively when they are in the jump pattern. Let variable m be the first time they meet on a specific position. Supposing J_1, J_2 rendezvous exactly $x \geq 1$ times, then:

$$Pr(m \leq i) = \frac{\binom{N+1-i}{x}}{\binom{N}{x}}, \forall 1 \leq i \leq N - x + 1. \qquad (15.26)$$

For any given N and fixed value $1 \leq x \leq N$, denote the expected time to rendezvous as $E(m, x)$ which can be formulated as:

$$
\begin{aligned}
E(m, x) &= \sum_{i=1}^{N-x+1} i \cdot Pr(m = i) \\
&= \sum_{i=1}^{N-x+1} Pr(m \leq i) \\
&= \sum_{i=1}^{N-x+1} \frac{\binom{N+1-i}{x}}{\binom{N}{x}} \\
&= \frac{N+1}{x+1}
\end{aligned}
\tag{15.27}
$$

We accumulate the expectations for all possible N and x to derive:

$$
\begin{aligned}
E(J, J) &= \sum_{x=0}^{N} p(N, x) \cdot E(m, x) \\
&= \sum_{x=1}^{N} p(N, x) \cdot E(m, x) + p(N, 0)(N + T_s)
\end{aligned}
\tag{15.28}
$$

here $p(N, x)$ is the probability that J_1, J_2 rendezvous exactly $x \geq 1$ times. On the basis that x rendezvous points exist between J_1, J_2, the remaining part cannot rendezvous and the probability is denoted as $p(N - x, 0)$. As the x rendezvous points have $x!$ different permutations, we derive:

$$
p(N, x) = \frac{p(N - x, 0)}{x!}
\tag{15.29}
$$

Combining Eq. (15.27), we get

$$
\begin{aligned}
\sum_{x=1}^{N} p(N, x) \cdot E(m, x) &= \sum_{x=1}^{n} E(m, x) \cdot \frac{p(N - x, 0)}{x!} \\
&= (N + 1) \cdot \sum_{x=1}^{n} \frac{p(N - x, 0)}{(x + 1)!} \\
&= (N + 1) \cdot \sum_{x=1}^{n} p(N + 1, x + 1) \\
&= (N + 1)(1 - p(N, 0) - p(N + 1, 0))
\end{aligned}
\tag{15.30}
$$

Plugging this into the formulation of $E(J, J)$, the lemma holds.

Then we need to calculate $p(N, 0)$ which denotes the probability that J_1, J_2 do not rendezvous. Assuming J_1 is the permutation generated by user A, we count the number of permutations (J_2) that do not rendezvous with J_1 (denote the number as D_N), which can be computed as in Lemma 15.5.

Lemma 15.5 $D_N = N! \cdot \sum_{k=0}^{N}(-1)^k \cdot \frac{1}{k!}$. When N is large enough, $D_N = \lfloor \frac{N!}{e} \rfloor$

Proof Consider two permutations J_1, J_2 of $[N]$ generated by user A and user B respectively. Let $J_1(i), J_2(i)$ be the labels of the i-th position. Since no rendezvous happens, $J_1(N)$ does not represent the same universal port as $J_2(N)$. Suppose $J_2(N)$ and $J_1(i)$, $1 \le i < N$ represent the same universal port while $J_1(N)$ and $J_2(j)$, $1 \le j < N$ represent the same universal port.

(1) If $i = j$, rendezvous cannot happen for all other $N - 2$ positions and the number of such permutations is D_{N-2};
(2) if $i \neq j$, the number of such permutations is D_{N-1}.

Therefore, we can compute:

$$D_N = (N - 1)(D_{N-1} + D_{N-2}) \tag{15.31}$$

It is easy to see $D_1 = 0, D_2 = 1$ and $p(N, 0) = \frac{D_N}{N!}$. Plugging these into the equation we get:

$$N!p(N, 0) = (N - 1)((N - 1)!p(N - 1, 0) + (N - 2)!p(N - 2, 0))$$

After the transformation we get:

$$N!p(N, 0) - N!p(N - 1, 0) = -(N - 1)!p(N - 1, 0) + (N - 1)!p(N - 2, 0) \tag{15.32}$$

Let $w(N) = N!p(N, 0) - N!p(N - 1, 0)$, we can solve the above equation as:

$$w(N) = -w(N - 1) = (-1)^{N-1}w(1) = (-1)^{(N-1)}$$

Then, we have:

$$p(N, 0) - p(N - 1, 0) = \frac{1}{N!}(-1)^N \tag{15.33}$$

and we can solve the equation as:

$$p(N, 0) = \frac{1}{N!} \sum_{i=0}^{N}(-1)^i \frac{1}{i!} \tag{15.34}$$

Therefore, D_N is computed as:

$$D_N = N! \cdot p(N, 0) = N! \cdot \sum_{k=0}^{N}(-1)^k \cdot \frac{1}{k!} \tag{15.35}$$

Fig. 15.1 An example of the overlapping between two random permutations

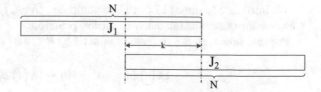

When $N \to \infty$, $p(N, 0)$ is the Taylor expansion of e^{-1}, and thus $D_N = \lfloor \frac{N!}{e} \rfloor$. So the lemma holds.

Since $p(N, 0) = \frac{D_N}{N!}$, we can combine Eqs. (15.20)–(15.25) to derive the expected time to rendezvous as in Theorem 15.2.

Theorem 15.2 *The expected time to rendezvous (ETTR) of the SRS algorithm (Algorithm 15.1) for two synchronous and port-symmetric users can be formulated as:*

$$T_s = \frac{T_1 + T_2 + T_3}{1 - p^2(1 - \frac{1}{N}) - (1 - p)^2 p(N, 0)} \tag{15.36}$$

where:

$$
\begin{aligned}
T_1 &= p(1 - p)(N + 1) \\
T_2 &= (1 - p)^2 [(N + 1)(1 - p(N, 0) - p(N + 1, 0)) + p(N, 0) \cdot N] \\
T_3 &= [p^2(\frac{1}{N} + N - 1)]
\end{aligned}
\tag{15.37}
$$

In order to find out the optimal p that minimizes T_s, let $\frac{dT_s}{dp} = 0$, and we can compute the value of p. When $N \to \infty$, $p \approx 0.2475$ and $T_s \approx 0.8289N$, which matches the state-of-the-art results [2].

15.2.4 Asynchronous Users Scenario

In order to analyze the algorithm for two asynchronous users, we present a method to derive the *ETTR* value for a general situation, i.e. an arbitrary N value. Similar to the analysis for two synchronous users, we first consider the scenario where two users are both in the jump pattern and are in the asynchronous situation.

Suppose sequences J_1, J_2 are two random permutations of $[N]$ generated by users A and B respectively. Let $r(N, k)$ denote the event that two users rendezvous in the overlapping fragment of length k (as in Fig. 15.1) and $R(N, k)$ denote the corresponding variable. Let $p(N, k, j)$ be the probability that they rendezvous exactly j times in the overlapping part; it is obvious that:

$$Pr(\overline{r(N, k)}) = p(N, k, 0) \tag{15.38}$$

We introduce Lemmas 15.6–15.8 to compute $p(N, k, j)$ and $E(R(N, k))$. To begin with, we introduce the inclusion-exclusion principle.

For two sets A, B, the cardinality of set $A \bigcup B$ can be computed as:

$$|A \bigcup B| = |A| + |B| - |A \bigcap B| \tag{15.39}$$

When there are multiple sets $A_1, A_2, \ldots, A_n$, we can compute the cardinality of set $\bigcup_{i=1}^{n} A_i$ as:

$$
\left| \bigcup_{i=1}^{n} \right| = \sum_{i=1}^{n} A_i - \sum_{1 \le i < j \le n} |A_i \bigcap A_j| + \sum_{1 \le i < j < k \le n} |A_i \bigcap A_j \bigcap A_k|
$$

$$
+ \cdots + (-1)^{n-1} |A_1 \bigcap A_2 \bigcap \cdots \bigcap A_n| \tag{15.40}
$$

$$
= \sum_{k=1}^{n} (-1)^{k+1} \left(\sum_{1 \le i_1 < \ldots < i_k \le n} |A_{i_1} \bigcap \cdots \bigcap A_{i_k}| \right)
$$

Lemma 15.6 $p(N, k, 0) = \sum_{i=0}^{k} (-1)^i \cdot \frac{\binom{k}{i}}{A_N^i}$.

Proof This lemma can be derived easily. Denote $q(N, k, i)$ as the probability that two users rendezvous at least i times in the overlapping fragment which has length k; when $1 \le i \le k$, we have:

$$q(N, k, i) = \frac{\binom{k}{i} \cdot (N - i)!}{N!} = \frac{\binom{k}{i}}{A_N^i} \tag{15.41}$$

Applying the inclusion-exclusion principle,

$$p(N, k, 0) = 1 - q(N, k, 1) + q(N, k, 2) + \cdots + (-1)^i q(N, k, i) = \sum_{i=0}^{k} (-1)^i \cdot \frac{\binom{k}{i}}{A_N^i} \tag{15.42}$$

so the lemma holds.

Lemma 15.7 $p(N, k, j) = p(N - j, k - j, 0) \cdot \frac{\binom{k}{j}}{A_N^j}$.

Proof Let $D(N, k, j)$ denote the number of permutations when J_1, J_2 have overlapping length k and exactly j rendezvous points. It is obvious that:

$$D(N, k, j) = N! \cdot N! \cdot p(N, k, j) \tag{15.43}$$

Similarly, we compute:

$$D(N - j, k - j, 0) = (N - j)! \cdot (N - j)! \cdot p(N - j, k - j, 0) \tag{15.44}$$

For any instance of the $D(N - j, k - j, 0)$ situations, it can be transformed into some instance in the $D(N, k, j)$ situations. Clearly, there are $\binom{N}{j}$ numbers (rendezvous points) that can be chosen, and there are $k - j + 1$ positions to place the first number, $k - j + 2$ positions for the second one, until $k - j + j$ positions for the j-th number. Thus, we derive:

$$D(N, k, j) = D(N - j, k - j, 0) \cdot \binom{N}{j} \cdot \frac{k!}{(k - j)!} \tag{15.45}$$

Combining the relationships of $D(N, k, j), p(N, k, j)$ and $D(N - j, k - j, 0), p(N - j, k - j, 0)$, we get:

$$
\begin{aligned}
p(N, k, j) &= \frac{D(N, k, j)}{N! \cdot N!} = \frac{D(N - j, k - j, 0) \cdot \binom{N}{j} \cdot \frac{k!}{(k-j)!}}{N! \cdot N!} \\
&= p(N - j, k - j, 0) \cdot \frac{(N - j)! \cdot (N - j)!}{N! \cdot N!} \cdot \binom{N}{j} \cdot \frac{k!}{(k - j)!} \\
&= p(N - j, k - j, 0) \cdot \frac{\binom{k}{j}}{A_N^j}
\end{aligned}
\tag{15.46}
$$

Thus the lemma holds.

Similar to Lemma 15.4, we bound the *ETTR* of $R(N, k)$ in Lemma 15.8.

Lemma 15.8 $E(R(N, k)) = (N + 1)(1 - p(N + 1, k + 1, 0)) - (k + 1) p(N, k, 0)$.

Proof When $j > k$, $p(N, k, j) = 0$ and we accumulate the probabilities when $j = 0, 1, \ldots, k$ as:

$$\sum_{j=0}^{k} p(N, k, j) = p(N, k, 0) + p(N, k, 1) + \cdots + p(N, k, k) = 1. \tag{15.47}$$

Supposing two users rendezvous exactly j times in the overlapping part of length k (denote the event as $r(N, k, j)$). Let $r_{k,j,1}$ be the time when they first rendezvous and let q_i be the probability that $r_{k,j,1}$ is no more than i, thus:

$$q_i = Pr(r_{k,j,1} \le i \mid r(N, k, j)) = \frac{\binom{k+1-i}{j}}{\binom{k}{i}} \tag{15.48}$$

where $i \le k + 1 - j$. When $i > k + 1 - j$, $q_i = 0$. We can formulate the expected time of the first rendezvous as:

$$E(r_{k,j,1} \mid r(N, k, j)) = \sum_{i=1}^{k+1-j} i \cdot Pr(r_{k,j,1} = i \mid r(N, k, j))$$

$$= \sum_{i=1}^{k+1-j} Pr(r_{k,j,1} \leq i \mid r(N, k, j)) \tag{15.49}$$

$$= \sum_{i=1}^{k+1-j} q_i = \frac{\sum_{i=1}^{k+1-j} \binom{k+1-i}{j}}{\binom{k}{j}}$$

$$= \frac{k+1}{j+1}$$

Thus we accumulate all the expectations when $j = 1, 2, \ldots, k$ as:

$$E(R(N, k)) = \sum_{j=1}^{k} p(N, k, j) \cdot E(r_{k,j,1} \mid r(N, k, j))$$

$$= \sum_{j=1}^{k} p(N - j, k - j, 0) \cdot \frac{\binom{k}{j}}{A_N^j} \cdot \frac{k+1}{j+1}$$

$$= (N + 1) \cdot \sum_{j=1}^{k} p(N - j, k - j, 0) \frac{\binom{k+1}{j+1}}{A_{N+1}^{j+1}} \tag{15.50}$$

$$= (N + 1) \cdot \sum_{j=1}^{k} p(N + 1, k + 1, j + 1)$$

$$= (N + 1)(1 - p(N + 1, k + 1, 0)) - (k + 1)p(N, k, 0)$$

We use Lemma 15.7 and plug in Eq. (15.47) to derive Eq. (15.50), and thus the lemma holds.

Without loss of generality, suppose user B starts the algorithm δ time slots later than user A. Since each user makes a choice every N time slots independently, we consider the situation as user B starts:

$$d = \delta \mod N \tag{15.51}$$

time slots later than user A. Let T_1 be the *ETTR* value when user A is in the stay pattern, and T_2 be the *ETTR* value when user A is in the jump pattern. Then we derive the *ETTR* value for the asynchronous scenario as follows.

Theorem 15.3 *For an arbitrary N, the optimal p of Algorithm 15.1 can be determined numerically and the minimized ETTR is computed as:*

$$ETTR = p \cdot T_1 + (1 - p) \cdot T_2 \tag{15.52}$$

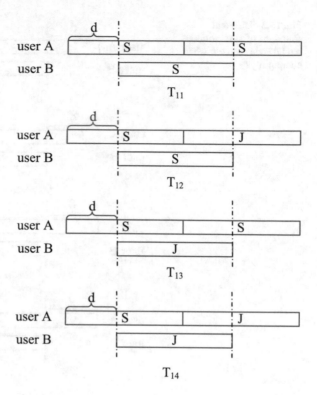

Fig. 15.2 Different situations of asynchronous rendezvous scenario when computing T_1

Proof In order to compute T_1, T_2, there are 4 situations respectively as shown in Figs. 15.2 and 15.3.

Since we treat every N time slots as a block, user B's first block intersects with user A's two consecutive blocks. Let B_1 denote user B's first block's pattern, and A_1, A_2 denote user A's two intersecting blocks' patterns. For simplicity, we write $B_1 = S$ for the stay pattern and $B_1 = J$ for the jump pattern. Thus:

$$T_1 = ETTR(A_1 = S)$$
$$T_2 = ETTR(A_1 = J) \tag{15.53}$$

Denote $A_1 \cap B_1$ and $A_2 \cap B_1$ as the overlapping fragment, and thus:

$$|A_1 \cap B_1| = N - d$$
$$|A_2 \cap B_1| = d \tag{15.54}$$

here $|.|$ represents the length of the overlapping part. The situations of the overlapping fragments are also illustrated in Figs. 15.2 and 15.3.

As depicted in Fig. 15.2, we denote

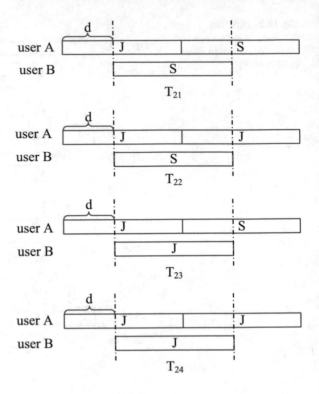

Fig. 15.3 Different situations of asynchronous rendezvous scenario when computing T_2

$$T_{11} = ETTR(B_1 = S, A_2 = S \mid A_1 = S)$$
$$T_{12} = ETTR(B_1 = S, A_2 = J \mid A_1 = S)$$
$$T_{13} = ETTR(B_1 = J, A_2 = S \mid A_1 = S)$$ \hfill (15.55)
$$T_{14} = ETTR(B_1 = J, A_2 = J \mid A_1 = S)$$

and we can get the formulation of T_1:

$$T_1 = p^2 \cdot T_{11} + p(1 - p) \cdot T_{12} + (1 - p)p \cdot T_{13} + (1 - p)^2 \cdot T_{14} \qquad (15.56)$$

Similarly, from Fig. 15.3, we derive Eq. (15.57) for T_2:

$$T_2 = p^2 \cdot T_{21} + p(1 - p) \cdot T_{22} + (1 - p)p \cdot T_{23} + (1 - p)^2 \cdot T_{24} \qquad (15.57)$$

where:

$$T_{21} = ETTR(B_1 = S, A_2 = S \mid A_1 = J)$$
$$T_{22} = ETTR(B_1 = S, A_2 = J \mid A_1 = J)$$
$$T_{23} = ETTR(B_1 = J, A_2 = S \mid A_1 = J)$$ \hfill (15.58)
$$T_{24} = ETTR(B_1 = J, A_2 = J \mid A_1 = J)$$

Now we present the method to produce the expressions of $T_{11}, T_{12}, T_{13}, T_{14}$ and $T_{21}, T_{22}, T_{23}, T_{24}$.

Let $r_1(T_{ij})$ and $r_2(T_{ij})$ be the events that the users rendezvous in $A_1 \bigcap B_1$, $A_2 \bigcap B_1$, respectively, and $R_1(T_{ij}), R_2(T_{ij})$ be the corresponding variables, where $1 \le i \le 2, 1 \le j \le 4$. We derive the formulation of T_{ij} as:

$$T_{ij} = Pr(\overline{r_1(T_{ij})})[(N - d) + Pr(r_2(T_{ij})) \cdot E(R_2(T_{ij})) + Pr(\overline{r_2(T_{ij})}) \cdot (d + T_\delta)$$
$$+ Pr(r_1(T_{ij})) \cdot E(R_1(T_{ij}))$$

(15.59)

where $\delta = (j - 1) \mod 2 + 1$ (i.e. $\delta = 1$ when $j = 1, 3$; otherwise $\delta = 2$). Thus we can plug in the following probabilities and expectations to generate Eqs. (15.60) and (15.61).

For event $r_1(T_{11})$, we compute:

$$\begin{cases} Pr(r_1(T_{11})) = \frac{1}{N} \\ E(R_1(T_{11})) = 1 \end{cases}$$

For event $r_2(T_{11})$, we compute:

$$\begin{cases} Pr(r_2(T_{11})) = \frac{1}{N} \\ E(R_2(T_{11})) = 1 \end{cases}$$

For event $r_1(T_{12})$, we compute:

$$\begin{cases} Pr(r_1(T_{12})) = \frac{1}{N} \\ E(R_1(T_{12})) = 1 \end{cases}$$

For event $r_2(T_{12})$, we compute:

$$\begin{cases} Pr(r_2(T_{12})) = \frac{d}{N} \\ E(R_2(T_{12})) = \frac{d+1}{2} \end{cases}$$

For event $r_1(T_{13})$, we compute:

$$\begin{cases} Pr(r_1(T_{13})) = \frac{N-d}{N} \\ E(R_1(T_{13})) = \frac{N-d+1}{2} \end{cases}$$

For event $r_2(T_{13})$, we compute:

$$\begin{cases} Pr(r_2(T_{13})) = \frac{d}{N} \\ E(R_2(T_{13})) = \frac{d+1}{2} \end{cases}$$

For event $r_1(T_{14})$, we compute:

$$\begin{cases} Pr(r_1(T_{14})) = \frac{N-d}{N} \\ E(R_1(T_{14})) = \frac{N-d+1}{2} \end{cases}$$

For event $r_2(T_{14})$, we compute:

$$\begin{cases} Pr(\overline{r_2(T_{14})}) = p(N, d, 0) \\ Pr(r_2(T_{14})) \cdot E(R_2(T_{11})) = E(R(N, d)) \end{cases}$$

For event $r_1(T_{21})$, we compute:

$$\begin{cases} Pr(r_1(T_{21})) = \frac{N-d}{N} \\ E(R_1(T_{21})) = \frac{N-d+1}{2} \end{cases}$$

For event $r_2(T_{21})$, we compute:

$$\begin{cases} Pr(r_2(T_{21})) = \frac{1}{N} \\ E(R_2(T_{21})) = 1 \end{cases}$$

For event $r_1(T_{22})$, we compute:

$$\begin{cases} Pr(r_1(T_{22})) = \frac{N-d}{N} \\ E(R_1(T_{21})) = \frac{N-d+1}{2} \end{cases}$$

For event $r_2(T_{22})$, we compute:

$$\begin{cases} Pr(r_2(T_{22})) = \frac{d}{N} \\ E(R_2(T_{22})) = \frac{d+1}{2} \end{cases}$$

For event $r_1(T_{23})$, we compute:

$$\begin{cases} Pr(\overline{r_1(T_{23})}) = p(N, N-d, 0) \\ Pr(r_1(T_{23})) \cdot E(R_1(T_{23})) = E(R(N, N-d)) \end{cases}$$

For event $r_2(T_{23})$, we compute:

$$\begin{cases} Pr(r_2(T_{23})) = \frac{d}{N} \\ E(R_2(T_{23})) = \frac{d+1}{2} \end{cases}$$

For event $r_1(T_{24})$, we compute:

Table 15.1 Optimal p and minimized *ETTR* values in Algorithm 15.1

N	Optimal p	ETTR	ETTR/N
3	0.302	2.887	0.9624
5	0.280	4.749	0.9499
10	0.233	9.332	0.9332
50	0.206	45.765	0.9159
100	0.203	91.354	0.9135
200	0.202	182.467	0.9123
500	0.201	455.806	0.9116
1000	0.200	911.369	0.9113
2000	0.200	1822.432	0.9112
10000	0.200	911.149	0.9111
$N \to \infty$	0.200	0.9111N	0.9111

$$\begin{cases} Pr(\overline{r_1(T_{24})}) = p(N, N - d, 0) \\ Pr(r_1(T_{24})) \cdot E(R_1(T_{24})) = E(R(N, N - d)) \end{cases}$$

For event $r_2(T_{24})$, we compute:

$$\begin{cases} Pr(\overline{r_2(T_{24})}) = p(N, d, 0) \\ Pr(r_2(T_{24})) \cdot E(R_2(T_{24})) = E(R(N, d)) \end{cases}$$

Combining these equations, we can derive the expressions of $T_{11}, T_{12}, T_{13}, T_{14}$, as follows:

$$\begin{cases} T_{11} = \frac{1}{N} \cdot 1 + \frac{N-1}{N} \cdot \frac{1}{N} \cdot (N - d + 1) + (\frac{N-1}{N})^2 \cdot (N + T_1) \\ T_{12} = \frac{1}{N} \cdot 1 + \frac{N-1}{N} \cdot \frac{d}{N} \cdot (N - d + \frac{d+1}{2}) + \frac{N-1}{N} \cdot \frac{N-d}{N} \cdot (N + T_2) \\ T_{13} = \frac{N-d}{N} \cdot \frac{N-d+1}{2} + \frac{d}{N} \cdot \frac{d}{N} \cdot (N - d + \frac{d+1}{2}) + \frac{d}{N} \cdot \frac{N-d}{N} \cdot (N + T_1) \\ T_{14} = \frac{N-d}{N} \cdot \frac{N-d+1}{2} + \frac{d}{N} \cdot (N - d + E(R(N, d))) + \frac{d}{N} \cdot p(N, d, 0) \cdot (d + T_2) \end{cases} \tag{15.60}$$

Similarly, we derive the expression of $T_{21}, T_{22}, T_{23}, T_{24}$:

$$\begin{cases} T_{21} = \frac{N-d}{N} \cdot \frac{N-d+1}{2} + \frac{d}{N} \cdot \frac{1}{N} \cdot (N - d + 1) + \frac{d}{N} \cdot \frac{N-1}{N} \cdot (N + T_1) \\ T_{22} = \frac{N-d}{N} \cdot \frac{N-d+1}{2} + \frac{d}{N} \cdot \frac{d}{N} \cdot (N - d + \frac{d+1}{2}) + \frac{d}{N} \cdot \frac{N-d}{N} \cdot (N + T_2) \\ T_{23} = E(R(N, N - d)) + p(N, N - d, 0) \cdot \frac{d}{N} \cdot (N - d + \frac{d+1}{2}) \\ \quad + p(N, N - d, 0) \cdot \frac{N-d}{N} \cdot (N + T_1) \\ T_{24} = E(R(N, N - d)) + p(N, N - d, 0)(N - d + E(R(N, d))) \\ \quad + p(N, N - d, 0)p(N, d, 0)(d + T_2) \end{cases} \tag{15.61}$$

Combining Eqs. (15.52)–(15.61), p is optimized numerically for arbitrary N and the minimized *ETTR* of our algorithm can be computed as Eq. (15.52). Table 15.1 lists some results derived through this numerical method.

15.3 Port-Asymmetric Rendezvous

In Sect. 15.2, we introduce a good method that works better than picking a random port for rendezvous when the users have symmetric available ports. In this section, we handle the port-asymmetric situations and present several randomized algorithms.

15.3.1 Random Picking Algorithm

One trivial way to handle the oblivious blind rendezvous between two anonymous, port-asymmetric users is to pick the available port for rendezvous randomly. We describe such an algorithm in Algorithm 15.2.

Algorithm 15.2 Random Picking Algorithm

1: Denote the set of the user's available port set as C;
2: Denote $C = \{c(1), c(2), \ldots, c(k)\}$ where $k = |C|$;
3: $t := 0$;
4: **while** Not terminated **do**
5:    Pick a random number $i \in [1, k]$ and access port $c(i)$ in time t;
6:    $t := t + 1$;
7: **end while**

As depicted in the algorithm, the user has k available ports and it labels these ports locally as:

$$\{c(1), c(2), \ldots, c(l)\} \tag{15.62}$$

where each port $c(i)$ corresponds to a global port, but the user does not know the relationship between them. We derive the time complexity of achieving rendezvous with high probability.

Consider any two neighboring users u_a and u_b, and suppose the corresponding available ports sets are

$$C_a = \{c_a(1), c_a(2), \ldots, c_a(k_a)\}$$
$$C_b = \{c_b(1), c_b(2), \ldots, c_b(k_b)\} \tag{15.63}$$

respectively, where $k_a = |C_a|$ and $k_b = |C_b|$ record the number of available ports. Since we study rendezvous between two port-asymmetric users, sets C_a, C_b can be different.

Denote $C_g = C_a \cap C_b$, which represents the set of common available ports between user u_a and user u_b. Notice that, two users have at least one common available port and $|C_g| \geq 1$. We derive below the expected time to rendezvous of the random picking algorithm.

Lemma 15.9 *The expected time to rendezvous of the random picking algorithm is* $ETTR = \frac{|C_a||C_b|}{|C_g|}$ *for two port-asymmetric users.*

Proof Let r_t be the event when both users access the same universal port (the local labels may be different) in the t-th time slot. Since both users access the port randomly, we analyze the probability as follows.

User u_a accesses each available port randomly, and the probability of accessing each port $c_a(i)$ is:

$$Pr_a(i) = \frac{1}{|C_a|} \tag{15.64}$$

Similarly, the probability of user u_b accessing each port $c_b(j)$ is:

$$Pr_b(j) = \frac{1}{|C_b|} \tag{15.65}$$

Therefore, the probability of user u_a accessing port $c_a(i)$ and user u_b accessing port $c_b(j)$ at the same time is:

$$Pr(u_a accesses c_a(i), u_b accesses c_b(j)) = \frac{1}{|C_a||C_b|} \tag{15.66}$$

As there are $|C_g|$ common available ports for both users, for each port $c_g(l) \in C_g$, there exist i, j such that:

$$\begin{cases} c_a(i) = c_g(l) \\ c_b(j) = c_g(l) \end{cases}$$

here "=" means they correspond to the same universal port. Therefore, there are $|C_g|$ situations where they may access the same port and the probability is:

$$Pr(r_t) = \frac{|C_g|}{|C_a||C_b|} \tag{15.67}$$

Since both users make decisions randomly and independently in each time slot, we can compute the *ETTR* value as:

$$ETTR = \sum_{t=1}^{\infty} t \cdot Pr(r_t)$$
$$= \sum_{t=1}^{\infty} t \cdot \left(1 - \frac{|C_g|}{|C_a||C_b|}\right)^{(t-1)} \cdot \frac{|C_g|}{|C_a||C_b|} \qquad (15.68)$$
$$= \frac{|C_a||C_b|}{|C_g|}$$

Therefore, the lemma holds.

15.3.2 Random Prime Selection and Sequential Accessing Algorithm

Though the random picking algorithm has short expected time to rendezvous, it cannot guarantee rendezvous within a bounded number of time slots with high probability. Actually, we can design another algorithm that guarantees rendezvous if a certain condition is satisfied.

Algorithm 15.3 Random Prime Selection and Sequential Accessing Algorithm

---
1: Denote the set of the user's available port set as C;
2: Denote $C = \{c(1), c(2), \dots, c(k)\}$ where $k = |C|$;
3: Choose a random prime number $p \in [k, 3k]$;
4: $t := 0$;
5: **while** Not terminated **do**
6:   $x := t \mod p$;
7:   $index := (x - 1) \mod k + 1$;
8:   Access port $c(index)$ for rendezvous;
9:   $t := t + 1$;
10: **end while**

As described in Algorithm 15.3, suppose the user has k available ports and it chooses a random prime number p in the range of $[k, 3k]$. After picking prime number p, the user accesses port sequentially by its local labels from 1 to p. However, p may be larger than k and we map the number in $[k + 1, p]$ to $[1, k]$ as in Line 7. For example, $k = 2$ and we choose $p = 3$, and the user accesses the ports as in Fig. 15.4.

For two users u_a and u_b, denote their available port sets as:

$$C_a = \{c_a(1), c_a(2), \dots, c_a(k_a)\}$$
$$C_b = \{c_b(1), c_b(2), \dots, c_b(k_b)\} \qquad (15.69)$$

Fig. 15.4 An example of Algorithm 15.3

Time	1	2	3	4	5	6	7	8	9
Sequence	c(1)	c(2)	c(3)	c(1)	c(2)	c(3)	c(1)	c(2)	c(3)
Port	c(1)	c(2)	c(1)	c(1)	c(2)	c(1)	c(1)	c(2)	c(1)

respectively, where $k_a = |C_a|$ and $k_b = |C_b|$ record the number of available ports. Denote the chosen prime numbers for two users as p_a and p_b. We show that they can rendezvous within $p_a p_b$ time slots for sure, if $p_a \neq p_b$.

Theorem 15.4 *Two port-asymmetric users (synchronous or asynchronous) can achieve rendezvous within $p_a p_b$ time slots under the situation that $p_a \neq p_b$.*

Proof Denote the port accessing sequences of user u_a and u_b as:

$$S_a = \{c_a(1), c_a(2), \ldots, c_a(p_a), c_a(1), c_a(2), \ldots, c_a(p_a), \ldots\} \tag{15.70}$$

and

$$S_b = \{c_b(1), c_b(2), \ldots, c_b(p_b), c_b(1), c_b(2), \ldots, c_b(p_b), \ldots\} \tag{15.71}$$

We do not consider the situation where some port in $(c_a(k_a), c_a(p_a)]$ may not exist. Suppose user u_a is δ time slots earlier than user u_b. Consider one common available port c_g between two users. Suppose it corresponds to port $c_a(i)$ of user u_a and port $c_b(j)$ of user u_b. Suppose both users can rendezvous on port c_g after user u_b starts t time slots; then we deduce that:

$$\begin{cases} t + \delta \mod p_a \equiv i \\ t \mod p_b \equiv j \end{cases}$$

According to the Chinese Remainder Theorem (see Chap. 9, Theorem 9.1), such value t must exist which satisfies both equations and $t \leq p_a p_b$. Therefore, two users can always achieve rendezvous no matter when they start.

However, if both users choose the same prime number, they may never rendezvous if they happen to miss the common available port. However, the probability of such a failure is small.

15.4 Chapter Summary

In this chapter, we study the oblivious blind rendezvous (OBR) problem for two anonymous users that are indistinguishable from each other.

In the beginning, we show an impossibility result that no deterministic algorithm can tackle the OBR-2 problem even when the users start the rendezvous process at the same time (i.e. synchronous users). Then, we propose a randomized distributed algorithm called Stay or Random Selection (SRS) for a special situation in which all ports are available for the users, which performs better than randomly accessing all ports. Finally, we present several randomized algorithms for port-asymmetric users on the basis of a random picking strategy.

When all N ports are available, two anonymous users adopting the random selection method have expected time to rendezvous (*ETTR*) in N time slots. We prove that the optimal strategy when two users can run asymmetric algorithms, i.e. different strategies, has $ETTR = \frac{N+1}{2}$ time slots, where one user accesses a fixed port and the other accesses the ports according to a random permutation of the N ports. The SRS algorithm combines both ideas: the user accesses a fixed port for N time slots with probability p or accesses the ports according to a random permutation of the N ports (with probability $1 - p$).

Although the description of SRS is simple, it is difficult to compute the appropriate p value that minimizes the expected time to rendezvous. In the chapter, we show the complicated analyses for both synchronous and asynchronous situations:

(1) For two synchronous users, the $ETTR$ is derived in Theorem 15.2 and the optimal value of p can be derived numerically. When $N \to \infty, p \approx 0.2475$ and $ETTR \approx 0.8289N$, which matches the state-of-the-art result [2];
(2) For two asynchronous users, the $ETTR$ is derived in Theorem 15.3. Some detailed parameters are listed in Table 15.1 and when $N \to \infty$, $p \approx 0.200$, $ETTR = 0.9111N$.

Therefore, the SRS algorithm works better than random selection, which is an elegant and surprising result. However, we cannot claim that SRS is the optimal algorithm and one future direction is to explore the optimal algorithm when two users should run a symmetric strategy. Moreover, when not all ports are available for the users, which should be more practical, we need to design efficient randomized distributed algorithms that have a good performance in the future.

For the port-asymmetric rendezvous setting, the random picking algorithm can achieve rendezvous in $ETTR = \frac{|C_a||C_b|}{|C_g|}$ time slots, where C_a, C_b represent the number of available ports of the two users, while C_g denotes the number of common available ports. We also present another algorithm called the Random Prime Selection and Sequential Accessing Algorithm, which has good performance and the failure probability of no rendezvous within $p_a p_b$ time slots is very low, where p_a, p_b are two chosen prime numbers in the algorithm.

References

1. Alpern, S., & Pikounis, M. (2000). The telephone coordination game. *Game Theory Application*, 5, 1–10.
2. Anderson, E. J., & Weber, R. R. (1990). The rendezvous problem on discrete locations. *Journal of Applied Probability, 28*, 839–851.
3. Crawford, V. P., & Haller, H. (1990). Learning how to cooperate: Optimal play in repeated coordination game. *Econometrica, 58*(3), 571–596.
4. Fan, J. (2009). Symmetric rendezvous problem with overlooking. Ph.D. thesis, University of Cambridge.
5. Weber, R. R. (2006). The optimal strategy for symmetric rendezvous search on three locations. arXiv:0906.5447v1.
6. Weber, R. (2009). The Anderson-Weber strategy is not optimal for symmetric rendezvous search on K_4. arXiv:0912.0670.
7. Weber, R. R. (2012). Optimal symmetric rendezvous search on three locations. *Mathematics of Operations Research, 37*, 111–122.

Chapter 16
Oblivious Blind Rendezvous for Multi-user Multihop CRN

Abstract In this chapter, we propose the distributed oblivious blind rendezvous algorithm for multiple users in a multi-hop distributed system. As described in Problem 11.2, the system consists of M users with distinct identifers (IDs) in $[1, \hat{M}]$, where $\hat{M} \geq M$ and $\hat{M} \leq N^c$ for a constant value c (for simplicity, we re-use notation M to mean $\hat{M}$). Any two users in the system are connected within D hops, which implies the network diameter is D. In Sect. 16.1, we describe the algorithm for multiple users in a multi-hop system, and the correctness is presented in Sect. 16.2. We summarize the chapter in Sect. 16.3.

16.1 Algorithm Description

We adopt the intuitive idea in [1–3] to extend the algorithms for OBR-2 to multiple users: once every two users achieve rendezvous on a common port successfully, they can exchange their local information over the established communication link such that the input of the OBR-2 algorithms can be synchronized; then they would generate the same hopping sequence afterwards. Assume there are four parameters (I, M, N, C) (I is the user's ID, M is the maximum value of the ID, N is the number of all ports, and C is the set of available ports) used for each user and we extend the MSH algorithm (Algorithm 13.5) in Chap. 13 to the multi-user multi-hop scenario.

Algorithm 16.1 Rendezvous Algorithm for Multiuser Multihop Scenario

---
1: *Input: I, M, N, C*;
2: **while** Not terminated **do**
3:    Run the MSH algorithm (Alg. 13.5) with input (I, M, N, C);
4:    **if** Rendezvous with user - (I', M, N, C') **then**
5:       $I := \min(I, I')$;
6:       $C := C \bigcap C'$;
7:       Synchronize labels of the ports in C according to the user with smaller I;
8:    **end if**
9: **end while**

© Springer Nature Singapore Pte Ltd. 2017
Z. Gu et al., *Rendezvous in Distributed Systems*,
DOI 10.1007/978-981-10-3680-4_16

As described in Algorithm 16.1, the user runs the MSH algorithm with local para-meters (I, M, N, C). Once rendezvous is achieved with another user with parameters (I', M, N, C'), they exchange their information and three operations are executed:

(1) Change I to be the smaller value between I, I'[1];
(2) change C to be the intersection of C and C';
(3) synchronize the labels for the available ports in C according to the user with smaller I value such that $\forall i \in [1, |C|], c(i) = c'(i)$;

After these three steps, the four parameters of the two users are the same and they access the ports with the same hopping sequence until the next rendezvous happens. We derive the correctness and time complexity in Theorem 16.1.

16.2   Correctness and Complexity

Theorem 16.1 *Algorithm 16.1 guarantees that all users can achieve rendezvous in* $MTTR = O(N^2 D \log_N M)$ *time slots, where D is the diameter of the system.*[2]

Proof The theorem can be concluded through induction, similar to Theorem 10.1. We show that all users can update to the same I value, the same set of available ports $\bigcap_{i=1}^m C_i$ and the same labels of the ports in $O(N^2 D \log_N M)$ time slots.

By adopting the same analysis in Theorem 10.1, all users can generate the same set of available ports $\bigcap_{i=1}^m C_i$ after $4lNP * D = O(N^2 D \log_N M)$ time slots when all users have begun the rendezvous process. We show that all users can update to the same I value through induction. Denote the user with the smallest ID value as r, and we show that any user r_i can update the I value as user r's in $4lNP * d(r, r_i)$ time slots where $d(r, r_i)$ represents the minimum number of hops separating them.

(1) When $r_i = r$, $d(r, r_i) = 0$, it is satisfied obviously.
(2) Suppose for any user r_k with $d(r, r_k) \le k$, it updates I as the ID value of user r in $4lNP * d(r, r_k)$ time slots. Consider user r_{k+1} with $d(r, r_{k+1}) = k + 1$ and suppose it is connected to some user r_k with $d(r, r_k) = k$ (it is easy to see that such user must exist); user r_k has already updated the I value to be the same as user r in $4lNP * k$ time slots and user r_{k+1} can update the value as that of user r_k in $4lNP$ time slots when they rendezvous on a some common available port in $4lNP$ time slots from Theorem 13.3. Therefore, user r_{k+1} can also update I to have the ID value of user r in $4lNP * (k + 1)$ time slots.

Combining these two aspects, where D is the network diameter, all users in the network can update the I value as that of user r in $4lNPD = O(N^2 D \log_N M)$ time slots (the process can be thought of as user r sends its ID value to all users within D hops). After another $4lNPD$ time slots, all users can synchronize the labels of

---

[1] It does not mean the user really changes the ID value, but it only changes the input of the rendezvous algorithm.
[2] l is defined as in Theorem 13.3.

the available ports as user r which has the smallest ID value. Therefore, the users in the network would hop through the ports according to the same sequence after $O(N^2 D \log_N M)$ time slots. So the theorem holds.

16.3 Chapter Summary

In this chapter, we extend oblivious blind rendezvous algorithms between two users to multiple users in a multihop distributed system. Similar to the non-oblivious blind rendezvous problem, every two neighboring users can rendezvous on a common available port and their local information can be synchronized. Then, they can repeat the rendezvous attempt until all users finally access the same available port.

Similar to the discussion about rendezvous process among multiple users in a multihop distributed system (Chap. 10), we assume all users in the network share some common available port, which is impractical. In the future, we will design efficient algorithms to implement the system based on the rendezvous process between every pair of neighboring users.

References

1. Chuang, I., Wu, H.-Y., Lee, K.-R.,& Kuo, Y.-H. (2013). alternate hop-and-wait channel rendezvous method for cognitive radio networks. In *INFOCOM*.
2. Gu, Z., Hua, Q.-S., Wang, Y., & Lau, F. C. M. (2013). Nearly optimal asynchronous blind rendezvous algorithm for cognitive radio networks. In *SECON*.
3. Liu, H., Lin, Z., Chu, X., & Leung, Y.-W. (2012). Jump-stay rendezvous algorithm for cognitive radio networks. *IEEE Transactions on Parallel and Distributed Systems, 23*(10), 1867–1881.

Part IV
Distributed Rendezvous Applications

Chapter 17
Rendezvous in Heterogeneous Cognitive Radio Networks

Abstract Rendezvous is a fundamental and important process in operating a distributed system, which can be applied in many distributed applications running on the system. In this chapter, we introduce the rendezvous process in a special type of cognitive radio network: *Heterogeneous Cognitive Radio Network (HCRN)* where different users have different capabilities to sense the licensed spectrum. Many elegant rendezvous algorithms have been proposed by constructing sequences based on the channels' labels [1, 3, 7, 8, 10] or their identifiers (IDs) [2, 4, 5], and rendezvous can be guaranteed in a short time based on the special hopping sequences constructed. However, they all assume the users have the capability to sense and access *all* the licensed channels, which is unrealistic when the number of channels (N) is very large and some wireless devices may only operate on a small fraction of the channels. Therefore, HCRN is proposed, in which the users may have different spectrum-sensing capabilities. We introduce the system model and formulate the problem in Sect. 17.1. Rendezvous algorithms for the fully available spectrum are presented in Sect. 17.2, and rendezvous algorithms for the partially available spectrum are introduced in Sect. 17.3. Finally, we summarize the chapter in Sect. 17.4.

© Springer Nature Singapore Pte Ltd. 2017
Z. Gu et al., *Rendezvous in Distributed Systems*,
DOI 10.1007/978-981-10-3680-4_17

17.1   Preliminaries

We first introduce the system model of heterogeneous cognitive radio network (HCRN) and its difference with traditional cognitive radio network. Then, we define the rendezvous problem in the context of HCRN and show the challenges of designing efficient rendezvous algorithms for this kind of CRN.

17.1.1   System Model

The licensed spectrum is assumed to be divided into N non-overlapping channels:

$$U = \{1, 2, \ldots, N\} \tag{17.1}$$

Each user (here we mean secondary users) is equipped with a cognitive radio to sense the licensed spectrum. We say a channel is *available* for the user if it is not occupied by any nearby primary users (PUs) who own these licensed channels. Actually, the users may have different spectrum sensing capabilities and suppose user i can sense a set of continuous channels:

$$C_i = \{c_x, c_{x+1}, \ldots, c_{x+k_i-1}\} \subseteq U \tag{17.2}$$

which is assumed in [11, 12], where c_x is the starting channel and $k_i = |C_i|$, $1 \leq x \leq N - k_i + 1$.

The channels in set C_i are either occupied by nearby PUs or available for the (secondary) user i. We denote:

$$V_i \subseteq C_i \tag{17.3}$$

as the set of all available channels after the spectrum sensing stage.

Time is also assumed to be divided into slots of equal length $2t$, where t is sufficient for establishing a communication link if the users access the same channel at the same time slot. According to the IEEE 802.22 [9], t is often set to be 10 ms. The intuitive idea of setting each time slot to be $2t$ is to ensure that an overlap of t exists for link establishment even when the users do not start their process at aligned time slots (the idea is similar to the blind rendezvous for CRN in Chap. 5; we omit the details here).

Considering two users u_a and u_b with different spectrum sensing capability sets C_a, C_b, and the corresponding available channel sets V_a, V_b, they can rendezvous on some common available channel if $V_a \cap V_b \neq \emptyset$, which implies their capability sets must intersect.

Here we study two scenarios: *fully available spectrum* and *partially available spectrum*.

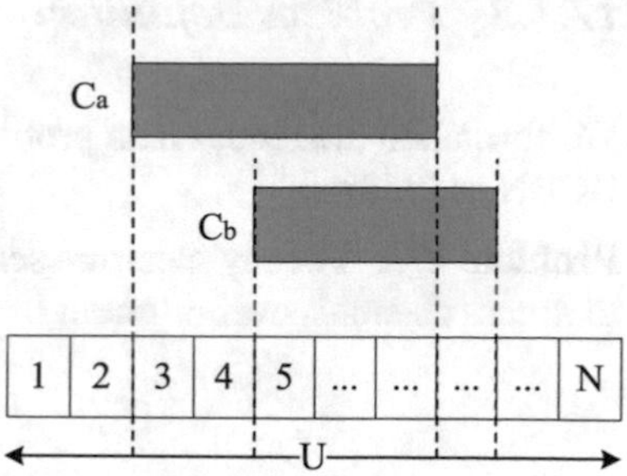

Fig. 17.1 An example of
different spectrum sensing
capability sets of two users

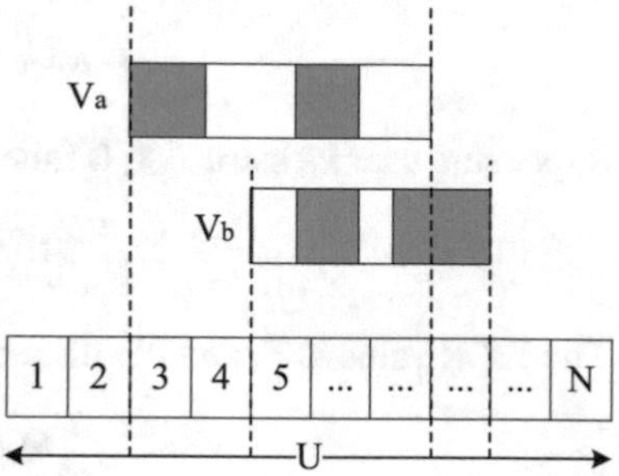

Fig. 17.2 An example of
different available channel
sets of two users

If all channels in the users' sensing capability sets are available after the spectrum sensing stage, we call that the fully available scenario (i.e. $V_i = C_i$). But in most circumstances, some channels are likely occupied ($V_i \neq C_i$) and we call that the partially available scenario.

For example, in Fig. 17.1, two users u_a and u_b have different sets of sensing capabilities $C_a, C_b \subseteq U$. If some channels are occupied by some PUs, we label these channels as white in Fig. 17.2 and the figure shows an example that two users have different sets of available channels, $V_a \subseteq C_a$ and $V_b \subseteq C_b$ respectively.

In Fig. 17.1, all channels in the user's sensing capability set are available and it is a fully available scenario, while Fig. 17.2 is a partially available scenario since some channels are occupied by the PUs [6].

In each time slot, user u_i can access an available channel from set V_i and attempt rendezvous with its potential neighbors. We say *rendezvous* happens when the users choose the same channel in the same time slot.

Time to rendezvous (*TTR*) denotes the number of time slots they take to rendezvous once all users have begun their attempt. Since the users are dispersed in different places and they may begin the rendezvous process in different time slots, we focus on designing efficient distributed algorithms for asynchronous users. We also use *Maximum Time to Rendezvous* (*MTTR*) to judge the performance of the rendezvous algorithms with respect to the worst situation.

17.1.2 Problem Definition

We formulate the rendezvous problem for the fully available spectrum scenario in HCRN as follows:

Problem 17.1 For any spectrum sensing capability set $C_i \subseteq U$, design an algorithm to access channels over different time slots:

$$t : f_{C_i}(t) \in C_i \tag{17.4}$$

such that for any two users u_a and u_b with sets:

$$C_a, C_b \subseteq U, C_a \cap C_b \neq \emptyset \tag{17.5}$$

Supposing user u_a starts $\delta \geq 0$ time slots earlier than user u_b,

$$\exists T_\delta, \text{ s.t. } f_{C_a}(T_\delta + \delta) = f_{C_b}(T_\delta) \tag{17.6}$$

The TTR value is T_δ and the maximum time to rendezvous is defined as:

$$MTTR = \max_{\forall \delta} T_\delta \tag{17.7}$$

The goal is to design rendezvous algorithms with bounded $MTTR$.

Although the fully available spectrum scenario rarely happens in practice, it represents the best spectrum condition that may happen in designing rendezvous algorithms for HCRN. For more general situations, we formulate the rendezvous problem for the partially available spectrum scenario as follows:

Problem 17.2 For any spectrum sensing capability set $C_i \subseteq U$ and available channel set $V_i \subseteq C_i$, design an algorithm to access channels over different time slots:

$$t : f_{C_i, V_i}(t) \in V_i \tag{17.8}$$

such that for any two users u_a and u_b with:

$$C_a, C_b \subseteq U, V_a \subseteq C_a, V_b \subseteq C_b, V_a \cap V_b \neq \emptyset \tag{17.9}$$

Supposing user u_a starts $\delta \geq 0$ time slots earlier than user u_b,

$$\exists T_\delta, \text{ s.t. } f_{C_a, V_a}(T_\delta + \delta) = f_{C_b, V_b}(T_\delta) \tag{17.10}$$

The TTR value is T_δ and the maximum time to rendezvous is defined as:

$$MTTR = \max_{\forall \delta} T_\delta \tag{17.11}$$

The goal is to design rendezvous algorithms with bounded $MTTR$.

For example, $U = \{1, 2, \ldots, 100\}$, and two capabilities sets are:

$$\begin{cases} C_a = \{2, 3, 4, 5, 6\} \\ C_b = \{5, 6, 7\} \end{cases}$$

Suppose that both users u_a and u_b adopt a simple algorithm by repeating the channels in their sensing capability set and user u_a is 1 time slot earlier than user u_b. As depicted in Fig. 17.3, they rendezvous on channel 5 at time slot 9, and thus $TTR = 14 - 1 = 13$ time slots. In fact, if the users apply the extant algorithms based on all channels in U, the maximum rendezvous time could be $O(N^2) \approx 10,000$ time slots, which is unacceptable. This figure is a simple example of the fully available spectrum scenario. When some channels are occupied, for example:

$$\begin{cases} V_a = \{2, 5, 6\} \\ V_b = \{6, 7\} \end{cases}$$

They cannot rendezvous on channel 5 and one more time slot is needed, as illustrated in Fig. 17.4. This is an example of the partially fully available spectrum scenario.

Time	1	2	3	4	5	6	7	8	9	10	11	12	13	14	15	16
user u_a	2	3	4	5	6	2	3	4	5	6	2	3	4	5	6	2
user u_b		5	6	7	5	6	7	5	6	7	5	6	7	5	6	7

Fig. 17.3 An example of rendezvous problem in HCRN

Time	1	2	3	4	5	6	7	8	9	10	11	12	13	14	15	16
user u_a	2	-	-	5	6	2	-	-	5	6	2	-	-	5	6	2
user u_b		-	6	7	-	6	7	-	6	7	-	6	7	-	6	7

Fig. 17.4 An example of rendezvous problem in HCRN when two users have partial available channels

Table 17.1 $MTTR$ comparison for fully and partially available scenarios in HCRN

Algorithms	Fully available scenario	Partially available scenario
HH [12]	$O(|C_A||C_B|)$	–
ICH [11]	$O(|C_A||C_B|)$	$O(|C_A||C_B|)$
TP [6]	$O(\max\{|C_A|, |C_B|\} \log \log N)$	–
MTP [6]	$O((\max\{|V_A|, |V_B|\})^2 \log \log N)$	$O((\max\{|V_A|, |V_B|\})^2 \log \log N)$

Remarks: (1) "–" means the algorithm is not applicable to the partially available spectrum scenario; (2) C_A, $C_B \subseteq U$ represent the capability sets of user A and B respectively; (3) $V_A \subseteq C_A$, $V_B \subseteq C_B$ represent the available channel sets of users A and B respectively

17.1.3   Challenges

In handling the blind rendezvous problem in HCRN, there are the following three *challenges*:

(1) First, different users may have different capabilities to sense the licensed spectrum, we should design efficient algorithms under such heterogeneity.
(2) Second, the users may start the rendezvous process at different time slots, and the rendezvous algorithms should work for both synchronous and asynchronous users with bounded rendezvous time.
(3) Third, traditional rendezvous algorithms have maximum time to rendezvous $(MTTR)$ as $MTTR = O(N^2)$, which is large when the user can only sense a small fraction of the channels. Thus, we should reduce the $MTTR$ value and guarantee fast rendezvous even for the worst situations see the comparison in Table 17.1.

17.2   Rendezvous for Fully Available Spectrum

In this section, we propose a new method called the *Traversing Pointer (TP)* algorithm for the users that have fully available channels. The intuitive idea is to accelerate the rendezvous process by accessing two channels at the same time, where one channel is fixed to be the first channel in the capability set, and the other is generated by hopping among the channels in the capability set. The method of generating such hopping sequence is similar to the method of time division in Chap. 7.

In the first place, we present a special construction for two available channels such that rendezvous can be guaranteed in $O(\log \log N)$ time slots if both users have only two available channels. Then, we introduce the TP algorithm on the basis of the special construction, which guarantees rendezvous for the two users with a fully available spectrum in a short time.

17.2.1 Rendezvous Scheme for Two Available Channels

Suppose each user has only two available channels, i.e. $|V_a| = |V_b| = 2$, and there exists at least one common channel, i.e. $V_a \cap V_b \neq \emptyset$. We present a special rendezvous scheme for the special scenario, which constructs a sequence of length $T_2 = 16(\lceil \log \log n \rceil + 1)$. The construction is based on three Disjoint Relaxed Difference Sets (DRDSs).

Supposing the available channel set of the user is:

$$V = \{v_1, v_2\} \subseteq U, \; where \; v_1 < v_2 \tag{17.12}$$

the method is described in Algorithm 17.1.

Algorithm 17.1 Rendezvous Scheme for Two Channels

1: $l_1 = \lceil \log N \rceil + 1, l_2 = \lceil \log l_1 \rceil + 1$;
2: Find the smallest number $c \in [1, l_1]$ such that the c-th bit of v_2 is 1 and the c-th bit of v_1 is 0;
3: Let $\overrightarrow{D} = \{*, c_{l_2}, c_{l_2-1}, \dots, c_1\}$ where $(c_{l_2}, c_{l_2-1}, \dots, c_1)$ is the binary representation of c;
4: Denote the rendezvous sequence $S = \emptyset$;
5: **for** $r = 1 : l_2 + 1$ **do**
6:   If $\overrightarrow{D}(r) = *$, add $S_* = (v_1, v_1, v_2, v_1, v_1, v_2, v_2, v_2)$ twice to S;
7:   If $\overrightarrow{D}(r) = 0$, add $S_0 = (v_1, v_1, v_2, v_1, v_2, v_1, v_2, v_2)$ twice to S;
8:   If $\overrightarrow{D}(r) = 1$, add $S_1 = (v_1, v_1, v_2, v_1, v_2, v_2, v_2, v_1)$ twice to S;
9: **end for**
10: Repeat the rendezvous sequence S until rendezvous;

Algorithm 17.1 finds the smallest number $c \in [1, l_1]$ such that the c-th bit of v_2 is 1 but the c-th bit of v_1 is 0, where $l_1 = \lceil \log N \rceil + 1$. Since $v_1 < v_2$, c must exist. It is obvious that c can be represented by $l_2 = \lceil \log \log N \rceil + 1$ binary bits. We construct vector $\overrightarrow{D}$ by adding a special symbol $*$ to the binary representation as in Line 3, and we construct the rendezvous sequence in $l_2 + 1$ rounds. In each round, different sequences S_*, S_0, S_1 are added twice to S and the intuitive idea of designing these sequences comes from the good properties of DRDS (see Definition 8.3 in Chap. 8). We define three sets as:

$$\begin{cases} D_* = \{\{1, 2, 4, 5\}, \{3, 6, 7, 8\}\} \\ D_0 = \{\{1, 2, 4, 6\}, \{3, 5, 7, 8\}\} \\ D_1 = \{\{1, 2, 4, 8\}, \{3, 5, 6, 7\}\} \end{cases}$$

It is easy to check that they are three DRDS under Z_8. S_*, S_0 and S_1 are then constructed on the basis of D_*, D_0, D_1 respectively. We show the construction of sequences S_0, S_1, S_* in Figs. 17.5, 17.6 and 17.7.

In each round, sequence S_*, S_0 or S_1 is added twice to the rendezvous sequence because the users can start the algorithm asynchronously. We first derive a useful lemma, as follows.

Fig. 17.5 Construction of sequence S_* on the basis of D_*

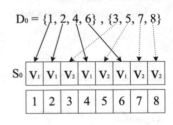

Fig. 17.6 Construction of sequence S_0 on the basis of D_0

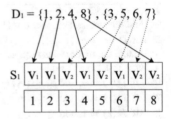

Fig. 17.7 Construction of sequence S_1 on the basis of D_1

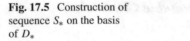

Lemma 17.1 *Every 8 continuous time slots in each round corresponds to a DRDS.*

Proof Consider the round containing two S_0 sequences where S_0 is constructed based on the DRDS D_0. Every 8 continuous time slots $[i, i + 7]$ where $1 \leq i \leq 9$ can be seen as rotating S_0 by $i - 1$ time slots. From the definition of Relaxed Difference Set (RDS) in Chap. 8, the rotation of an RDS is also an RDS. Thus the rotation of S_0 also corresponds to a DRDS. For example, when $i = 3$, the 8 continuous time slots are:

$$\{v_2, v_1, v_2, v_1, v_2, v_2, v_1, v_1\} \tag{17.13}$$

and they correspond to the DRDS:

$$\{\{2, 4, 7, 8\}, \{1, 3, 5, 6\}\} \tag{17.14}$$

We can also derive the same result for the other two sequences S_1, S_*, and thus the lemma holds.

Consider two users u_a and u_b with available channel sets:

$$\begin{cases} V_a = \{a_1, a_2\} \\ V_b = \{b_1, b_2\} \end{cases}$$

Suppose the chosen numbers in Line 2 are c_a, c_b respectively. We show the correctness of Algorithm 17.1 based on different relationships between c_a, c_b:

(1) If $c_a = c_b$, rendezvous is guaranteed in 16 time slots as in Lemma 17.2.
(2) If $c_a \neq c_b$, rendezvous is guaranteed in $16(\lceil \log \log N \rceil + 1)$ time slots as in Lemma 17.3.

Lemma 17.2 *Algorithm 17.1 guarantees rendezvous in 16 time slots if $c_a = c_b$.*

Proof When $c_a = c_b$, we claim that:

$$a_1 \neq b_2 \text{ and } a_2 \neq b_1 \tag{17.15}$$

If $a_1 = b_2$, we can derive:

$$b_1 < b_2 = a_1 < a_2 \tag{17.16}$$

From Line 2, the c_b-th bit of b_2 is 1 and the c_a-th bit of a_1 is 0, but $c_a = c_b$, which leads to a contradiction. Thus $a_1 \neq b_2$. Similarly, $a_2 \neq b_1$.

Since the users have at least one common channel, that is:

$$a_1 = b_1 \text{ or } a_2 = b_2 \tag{17.17}$$

We show that both pairs (a_1, b_1), (a_2, b_2) appear in the constructed sequences when two users have begun their process.

Denote the constructed sequences for the users as S_a and S_b respectively, and they are composed of $l_2 + 1$ rounds. We say the i-th round of user u_a (denoted as $r(a, i)$) overlaps with the j-th round of user u_b ($r(b, j)$) if their intersection length is at least 8 (time slots).

Without loss of generality, suppose user u_a is δ time slots earlier than user u_b. We show the lemma from two situations:

(1) If $r(b, 1)$ overlaps with $r(a, 1)$ and there are at least 8 overlapping time slots. By Lemma 17.1, the continuous 8 time slots correspond to two DRDSs for users u_a and u_b. From the definition of the DRDS, we can check that (a_1, b_1) and (a_2, b_2) both exist in the 8 time slots, and thus they rendezvous in the first round of user u_b.
(2) If $r(b, 1)$ overlaps with $r(a, i)$ where $1 < i \leq l_2 + 1$ and there are at least 8 overlapping time slots. If (a_1, b_1) does not exist in the intersecting 8 slots, channel b_1 meets a_2 in four time slots and b_2 also has to meet a_1 in four time slots. However, the sequence added in $r(b, 1)$ is different from the sequence in $r(a, i)$. Actually, S_* is added twice in $r(b, 1)$ while S_0 or S_1 is added in $r(a, i)$, and this situation cannot happen. Thus, (a_1, b_1) exists in the first round of user u_b. Similarly, we can prove that (a_2, b_2) exists. Thus they can rendezvous in 16 time slots.

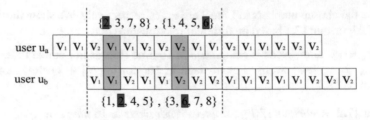

Fig. 17.8 An example of $r(b, 1)$ overlapping with $r(a, 1)$ in Algorithm 17.1

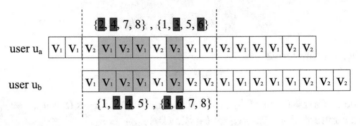

Fig. 17.9 An example of $r(b, 1)$ overlapping with $r(a, i)$ where $1 < i \le l_2 + 1$ in Algorithm 17.1

As depicted in Fig. 17.8, $r(b, 1)$ overlaps with $r(a, 1)$ and the first 8 overlapping time slots form two DRDSs are:

$$\begin{cases} \{\{2, 3, 7, 8\}, \{1, 4, 5, 6\}\} \text{ for user } u_a \\ \{\{1, 2, 4, 5\}, \{3, 6, 7, 8\}\} \text{ for user } u_b \end{cases}$$

Then we can check that (a_1, b_1) exists in the 2-nd time slot and (a_2, b_2) happens in the 6-th time slot. Similarly, Fig. 17.9 shows the example that $r(b, 1)$ overlaps with $r(a, i)$ where $1 < i \le l_2 + 1$, and both pairs (a_1, b_1) and (a_2, b_2) exist in the first overlapping 8 time slots. Therefore, the lemma holds.

Lemma 17.3 *Algorithm 17.1 guarantees rendezvous in* $T_2 = 16(\lceil \log \log N \rceil + 1)$ *time slots if* $c_a \ne c_b$.

Proof When $c_a \ne c_b$, there are four possible combinations of rendezvous situations:

$$\begin{cases} a_1 = b_1 \\ a_1 = b_2 \\ a_2 = b_1 \\ a_2 = b_2 \end{cases}$$

Thus the two users' overlapping sequences must contain the four pairs (a_1, b_1), (a_1, b_2), (a_2, b_1), (a_2, b_2). We show the lemma from two situations.

(1) If $r(b, 1)$ overlaps with $r(a, 1)$, (a_1, b_1), (a_2, b_2) exists in the overlapping part by Lemma 17.2. Since $c_a \neq c_b$, without loss of generality, suppose $c_a < c_b$ and there exists $1 \leq i \leq l_2$ such that the i-th bit of c_a is 0 but the i-th bit of c_b is 1 (such i must exist). When $r(b, i + 1)$ overlaps with $r(a, i + 1)$, we claim that (a_1, b_2) and (a_2, b_1) exist in the overlapping part. If (a_1, b_2) does not happen, a_1 has to meet b_1 four times and a_2 has to meet b_2 four times; however, $r(a, i + 1)$ and $r(b, i + 1)$ use different sequences (S_0 and S_1) and this situation cannot happen. Thus (a_1, b_2) appears at least once during the intersecting part. Similarly, (a_2, b_1) also exists. Therefore, rendezvous can be guaranteed in $16(i+1) \leq T_2$ time slots.

(2) If $r(b, 1)$ intersects with $r(a, i)$ where $1 < i \leq l_2 + 1$, the pairs (a_1, b_1) and (a_2, b_2) both exist by Lemma 17.2. Using the similar technique as the first situation, we can check that (a_1, b_2) and (a_2, b_1) exist in the first round of user u_b.

Combining the two situations, rendezvous can be guaranteed in T_2 time slots, and the lemma holds.

By Lemmas 17.2 and 17.3, we conclude the theorem:

Theorem 17.1 *Algorithm 17.1 guarantees rendezvous in* $T_2 = 16(\lceil \log \log n \rceil + 1)$ *time slots for the special situation that each user has two available channels.*

17.2.2 Traversing Pointer Algorithm

For the fully available spectrum scenario, we propose the Traversing Pointer (TP) algorithm based on the rendezvous scheme for two channels. Consider two users u_a and u_b with spectrum sensing capability sets $C_a, C_b \subseteq U$, the TP algorithm is described as Algorithm 17.2.

Algorithm 17.2 Traversing Pointer Algorithm

1: $t := 1, r := 1, L := 2T_2$;
2: $fp := c_x, mp := c_{x+k_i-1}$;
3: **while** not rendezvous **do**
4:     $r := \lfloor t/L \rfloor + 1, p := (t - 1)\%L + 1$;
5:     $r' := (r - 1)\%(2(k_i - 1))$;
6:     **if** $0 \leq r' < k_i - 1$ **then**
7:         $mp := c_{x+k_i-1-r'}$;
8:     **else**
9:         $mp := c_{x+r'\%(k_i-1)}$;
10:    **end if**
11:    Invoke Algorithm 17.1 with available channels $\{fp, mp\}$ and repeat the output twice to construct the rendezvous sequence $RS_r = \{s_1, s_2, \ldots, s_L\}$;
12:    Access the p-th channel of the sequence $s_p \in RS_r$;
13:    $t := t + 1$;
14: **end while**

Fig. 17.10 An illustration of Algorithm 17.2. fp is fixed at the first channel in all rounds, while mp traverses the channels back and forth and round by round

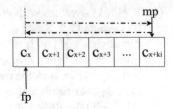

To begin with, suppose user i has the spectrum sensing capability set as:

$$C_i = \{c_x, c_{x+1}, \ldots, c_{x+k_i-1}\} \subseteq U \tag{17.18}$$

where $k_i = |C_i|$, $1 \le x \le N - k_i + 1$ and $\forall c_j \in C_i$, channel c_j is available.

The TP algorithm works on the basis of the rendezvous scheme for two available channels. There are two constructed 'pointers' where fp, i.e. *fixed pointer*, is fixed at the first channel c_x and mp is a *moving pointer* that traverses the capability set back and forth. We divide the time into rounds where each round contains $L = 2T_2$ time slots (we repeat the constructed sequence from Algorithm 17.1 twice to tackle the asynchronous situation). fp is fixed but mp changes in each round. As illustrated in Fig. 17.10, mp moves from the last channel c_{x+k_i-1} to the first one c_x in the first $k_i - 1$ rounds, and then from the first one to the last one in the next $k_i - 1$ rounds. The user continues the process until rendezvous.

17.2.3 Correctness and Complexity

Consider any two users u_a and u_b with spectrum sensing capability sets:

$$\begin{cases} C_a = \{c_x, c_{x+1}, \ldots, c_{x+k_a-1}\} \\ C_b = \{c_y, c_{y+1}, \ldots, c_{y+k_b-1}\} \end{cases}$$

where $1 \le x \le N - k_a + 1$, $1 \le y \le N - k_b + 1$. $C_a \cap C_b \ne \emptyset$ implies the following situation must happen:

$$c_x \in C_b \, or \, c_y \in C_a \tag{17.19}$$

Therefore, the constructed two pointers can help guarantee rendezvous when one user's moving pointer coincides with the other's fixed pointer. We derive the time complexity to achieve rendezvous in Theorem 17.2.

Theorem 17.2 *The TP algorithm (Algorithm 17.2) guarantees rendezvous for the fully available spectrum scenario in $O(\max\{|C_a|, |C_b|\} \log \log N)$ time slots.*

Proof Since the channels in the capability sets C_a and C_b are continuous and $C_a \cap C_b \ne \emptyset$, the first channel of C_a is in C_b (i.e. $c_x \in C_b$) or the first channel of C_b is in C_a (i.e. $c_y \in C_a$). Without loss of generality, suppose $c_x \in C_b$.

Denote the consecutive L time slots constructed in Line 11 as a round, and the chosen available channels in the r-th round of two users are $\{fp_{a,r}, mp_{a,r}\}, \{fp_{b,r}, mp_{b,r}\}$ respectively.

We say the i-th round of user u_a (denoted as $r_{a,i}$) overlaps with the j-th round of user u_b ($r_{b,j}$) if their intersection part contains at least $L/2$ time slots. From Theorem 17.1, if $r_{a,i}$ overlaps with $r_{b,j}$ and $\{fp_{a,i}, mp_{a,i}\} \cap \{fp_{b,j}, mp_{b,j}\} \neq \emptyset$, two users can achieve rendezvous in $L = 32(\lceil \log \log N \rceil + 1)$ time slots. There are two different situations according to the start time of two users:

(1) If user u_a starts earlier (no later) than user u_b, suppose the i-th round of user u_a overlaps with the first round of user u_b. We can find that, after $r = y + k_b - 1 - x$ rounds, $r_{a,i+r}$ overlaps with $r_{b,1+r}$ where user u_b's moving pointer chooses channel:

$$mp_{b,1+r} = c_{y+k_b-(1+r)} = c_x = fp_{a,i+r} \qquad (17.20)$$

thus, rendezvous is guaranteed in $(r + 1)L \leq |C_b|L$ time slots.

(2) If user u_b starts earlier than user u_a, suppose the i-th round of user u_b overlaps with the first round of user u_a, there are two situations according to the moving direction of user u_b's moving pointer (mp). It is easy to check that user u_b's moving pointer chooses channel c_x within $2k_b$ rounds no matter which direction it is heading. We omit the details and the reader may deduce the complexity of the situation. Therefore, rendezvous is guaranteed in $2|C_b|L$ time slots.

Similarly, when $c_y \in C_a$, rendezvous is also guaranteed in $2|C_a|L$ time slots. Therefore, the TP algorithm (Algorithm 17.2) guarantees rendezvous in $2 \max\{|C_a|, |C_b|\}L = O(\max\{|C_a|, |C_b|\} \log \log N)$ time slots when the spectrum is fully available.

In order to show the efficiency of the TP algorithm, we show a constructive lower bound in Theorem 17.3.

Theorem 17.3 $max\{|C_a|, |C_b|\}$ *time slots are needed to guarantee rendezvous for the fully available spectrum condition.*

Proof Suppose user u_a can sense only 1 channel (i.e. $|C_a| = 1$) which belongs to C_b. In order to discover the channel for rendezvous, user u_b has to traverse all channels in C_b at least once and thus (at least) $\max\{|C_a|, |C_b|\}$ time slots are needed, which concludes the theorem.

It is clear that the lower bound still holds if two users are synchronous, and the TP algorithm is nearly optimal with only an additional $O(\log \log N)$ factor. Compared with the state-of-the-art result $O(|C_a||C_b|)$ in [12], the TP algorithm removes an $O(\min\{|C_a|, |C_b|\})$ factor and it works more efficiently.

17.3   Rendezvous for Partially Available Spectrum

In this section, we propose the *Moving Traversing Pointer (MTP)* algorithm for the
users that have partially available channels. The intuitive idea is also to accelerate
the rendezvous process by accessing two channels at the same time, and the time
to rendezvous is only impacted by an $O(\log \log N)$ factor. When it comes to the
partially available scenario, the TP algorithm cannot work because the channel they
rendezvous on may be unavailable. Therefore, we propose this modified algorithm
where the 'fixed pointer' can also move after the 'moving pointer' has already traversed all channels in the capability set. Through such a modification, the MTP
algorithm can guarantee rendezvous in $O((\max\{|V_a|, |V_b|\})^2 \log \log N)$ time slots.

17.3.1   Moving Traversing Pointer Algorithm

In practical situations, the sensed available channels may be only a fraction of the
spectrum sensing capability set. For two users u_a and u_b with capability sets C_a, $C_b \subseteq$
U and available channel sets $V_a \subseteq C_a$, $V_b \subseteq C_b$, the TP algorithm may not guarantee
rendezvous.

For example, suppose the first channel of user u_a (c_x) belongs to user u_b's capability set ($c_x \in C_b$), c_x is available for user u_a ($c_x \in V_a$), but it is not available for
user u_b ($c_x \notin V_b$). The fixed pointer of user u_a stays at channel c_x all the time but
user u_b cannot access c_x, and thus rendezvous may not happen. In order to overcome
the disadvantage, we modify the TP algorithm and the intuitive idea is to move the
'fixed pointer' after the 'moving pointer' has already traversed the channels.

Similar to the assumption in the TP algorithm, suppose user i has the spectrum
sensing capability set as:

$$C_i = \{c_x, c_{x+1}, \ldots, c_{x+k_i-1}\} \subseteq U \tag{17.21}$$

where $k_i = |C_i|$ and $1 \le x \le n - k_i + 1$, and the available channel set is denoted as
$V_i \subseteq C_i$. Order the available channels by increasing order and denote:

$$V_i = \{c_{i,1}, c_{i,2}, \ldots, c_{i,m_i}\} \tag{17.22}$$

where $m_i = |V_i|$ and the equation:

$$\forall 1 \le j_1 < j_2 \le m_i, c_{i,j_1} < c_{i,j_2} \tag{17.23}$$

holds. The MTP algorithm is presented in Algorithm 17.3.

Algorithm 17.3 Moving Traversing Pointers Algorithm

1: $t := 1, r := 1, m_i = |V_i|$;
2: $L := 2T_2, P := 2(m_i - 1)L$;
3: $fp := c_{i,1}, mp := c_{i,m_i}$;
4: **while** Not rendezvous **do**
5: $l := \lfloor t/P \rfloor + 1, p_1 = (t-1)\%P + 1$;
6: $r := \lfloor p_1/L \rfloor + 1, p_2 := (p_1 - 1)\%L + 1$;
7: $l' := (l-1)\%m_i + 1, fp := c_{i,l'}$;
8: $r' := (r-1)\%(2(m_i - 1)) + 1$;
9: **if** $0 < r' < m_i$ **then**
10:      $mp := c_{i,m_i+1-r'}$;
11: **else**
12:      $mp := c_{i,r'\%(m_i-1)}$;
13: **end if**
14: Invoke Algorithm 17.1 with available channels $\{fp, mp\}$ and repeat the output twice to construct the rendezvous sequence $RS_{l,r} = \{s_1, s_2, \ldots, s_L\}$;
15: Access the p_2-th channel as $s_{p_2} \in RS_{l,r}$;
16: $t := t + 1$;
17: **end while**

The MTP algorithm (Algorithm 17.3) is different from the TP algorithm where the 'fixed pointer' does not always stay at the same channel. Assume time is divided into loops of length $P = 2(m_i - 1)L$ time slots and each loop contains $2(m_i - 1)$ rounds of length $L = 2T_2 = 32(\lceil \log \log N \rceil + 1)$. The pointer fp stays at a fixed available channel in each loop and it moves to the next available one every P time slots as in Line 7. Similar to the TP algorithm, the 'moving pointer' stays at a fixed channel in each round and traverses the available channels back and forth round by round. As illustrated in Fig. 17.11, fp is fixed at channel $c_{i,1}$ for the first P time slots and mp traverses from the last available channel c_{i,m_i} to the first one $c_{i,1}$, and then back to the last one every L time slots. In the next loop of P time slots, fp moves to channel $c_{i,2}$ as Fig. 17.12 and mp repeats the traversal. This process continues until rendezvous.

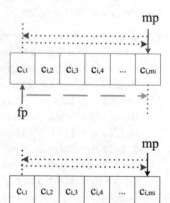

Fig. 17.11 mp traverses the channels back and forth and round by round, while fp moves to the next available channel every $2(m_i - 1)$ rounds

Fig. 17.12 mp traverses the channels back and forth and round by round, while fp moves to the next available channel every $2(m_i - 1)$ rounds

17.3.2 Correctness and Complexity

Consider users u_a and u_b with capability sets C_a, $C_b \subseteq U$ and available channel sets $V_a \subseteq C_a$, $V_b \subseteq C_b$ where $C_a \cap C_b \neq \emptyset$. Denote:

$$\begin{cases} V_a = \{c_{a,1}, c_{a,2}, \dots, c_{a,m_a}\} \\ V_b = \{c_{b,1}, c_{b,2}, \dots, c_{b,m_b}\} \end{cases}$$

where $m_a = |V_a|$, $m_b = |V_b|$. We show the correctness and the efficiency in the following theorem.

Theorem 17.4 *The MTP algorithm (Algorithm 17.3) guarantees rendezvous for the partially available spectrum scenario in* $O((\max\{|V_a|, |V_b|\})^2 \log \log N)$ *time slots.*

Proof Since $V_a \cap V_b \neq \emptyset$, there exist $1 \leq x \leq m_a, 1 \leq y \leq m_b$ such that one common available channel exists:

$$c_{a,x} = c_{b,y} \tag{17.24}$$

Denote the consecutive L time slots constructed in Line 14 as a round and every $2(m_i - 1)$ rounds as a loop ($i = a$ or b). Denote the r-th round of l-th loop for users u_a and u_b as $r_a(l, r)$ and $r_b(l, r)$, and the chosen available channels in the round as $\{fp_a(l, r), mp_a(l, r)\}$ and $\{fp_b(l, r), mp_b(l, r)\}$ respectively.

Similar to the analysis of Theorem 17.2, we say round $r_a(l_a, r_a)$ overlaps with round $r_b(l_b, r_b)$ if their intersection part contains at least $L/2$ time slots. From Theorem 17.1, if $r_a(l_a, r_a)$ overlaps with $r_b(l_b, r_b)$ and the chosen channels satisfy:

$$\{fp_a(l, r), mp_a(l, r)\} \cap \{fp_b(l, r), mp_b(l, r)\} \neq \emptyset \tag{17.25}$$

rendezvous can be achieved in the intersection part.

Without loss of generality, assuming $m_a = |V_a| \leq |V_b| = m_b$ and we show the theorem from two aspects.

(1) If user u_a starts the algorithm earlier than user u_b, suppose $r_a(l_a, r_a)$ overlaps with the first round of user u_b ($r_b(1, 1)$). After $(y - 1) \cdot 2(m_b - 1)$ rounds, user u_b's fixed pointer (fp) stays at channel $c_{b,y}$ for the next $2(m_b - 1)$ rounds. Since $2(m_b - 1) \geq 2(m_a - 1)$, user u_a's moving pointer (mp) has enough time (rounds) to traverse all available channels including $c_{a,x} = c_{b,y}$, and therefore the chosen channels overlap in $2(m_a - 1)$ rounds and rendezvous is guaranteed in $[2(m_b - 1) \cdot (y - 1) + 2(m_a - 1)] \cdot L \leq 2(m_b - 1)m_b L$ time slots.

(2) If user u_b starts the algorithm earlier than user u_a, suppose $r_b(l_b, r_b)$ overlaps with the first round of user u_a ($r_a(1, 1)$). Obviously, user u_b can get to the loop where the fixed pointer (fp) stays at channel $c_{b,y}$ in no more than $m_b - 1$ loops (i.e. $(m_b - 1) \cdot 2(m_b - 1)$ rounds). By the same analysis, rendezvous can be guaranteed in the following $2(m_a - 1)$ rounds, which implies the maximum rendezvous time can be bounded by:

$$[(m_b - 1) \cdot 2(m_b - 1) + 2(m_a - 1)] \cdot L \leq 2(m_b - 1)m_b L \qquad (17.26)$$

time slots.

Similarly, if $m_a \geq m_b$, we also can show that rendezvous is guaranteed in $2(m_a - 1)m_a L$ time slots. Therefore, Algorithm 17.3 guarantees rendezvous in:

$$2(\max\{m_a, m_b\})^2 \cdot 32(\lceil \log \log N \rceil + 1) = O((\max\{|V_a|, |V_b|\})^2 \log \log N) \quad (17.27)$$

time slots. Thus, the theorem holds.

17.4 Chapter Summary

The rendezvous problem has been widely studied in Cognitive Radio Networks (CRNs) since the unlicensed spectrum is overcrowded due to the increasing number of wireless devices, while the licensed spectrum is often underutilized. In this chapter, we introduce rendezvous processes in dealing with a special type of CRN where the users (such as the mobile phones or other wireless devices) can only detect a fraction of all channels. The different capabilities of detecting the licensed channels of the users create a *heterogeneous* network and this kind of network is called Heterogeneous Cognitive Radio Network (HCRN).

In the chapter, we study the simplest version of modeling the users' heterogeneous capabilities to detect the licensed channels, where each user can only sense a set of continuous channels. We mainly consider two scenarios, all channels in the users' sensing range are available, or part of them are available.

For the first situation, we introduce the Traversing Pointer (TP) algorithm, where two pointers exist to traverse the channels that the user can detect. This idea originates from the method of traversing the elements in an array, but it cannot work if the users' capability set is not continuous. For the second situation, we modify the TP algorithm and the proposed Moving Traversing Pointer (MTP) algorithm can traverse all channels in the users' capability set, while it keeps moving slowly to all available channels.

Rendezvous in an arbitrary HCRN can be more difficult if the users' capability set is discontinuous. The proposed "pointer" works well in the continuous capability set since we can regard it as an array, but we need to find out other efficient ways to handle more general heterogeneous capabilities.

References

1. Chen, S., Russell, A., Samanta, A. & Sundaram, R. (2014). Deterministic blind rendezvous in cognitive radio networks. In *ICDCS*.
2. Chuang, I., Wu, H. -Y., Lee, K. -R., & Kuo, Y. -H. (2013). Alternate hop-and-wait channel rendezvous method for cognitive radio networks. In *INFOCOM*.

3. Gu, Z., Hua, Q. -S., Wang, Y., & Lau, F. C. M. (2013). Nearly optimal asynchronous blind rendezvous algorithm for cognitive radio networks. In *SECON*.
4. Gu, Z., Hua, Q. -S. & Dai, W. (2014). Local sequence based rendezvous algorithms for cognitive radio networks. In *SECON*.
5. Gu, Z., Hua, Q. -S., & Dai, W. (2014). Fully distributed algorithms for blind rendezvous in cognitive radio networks. In *MOBIHOC*.
6. Gu, Z. Pu, H., Hua, Q. -S. & Lau, F. C. M. (2015). Improved rendezvous algorithms for heterogeneous cognitive radio networks. In *INFOCOM*.
7. Liu, H., Lin, Z., Chu, X., & Leung, Y. -W. (2012). Jump-stay rendezvous algorithm for cognitive radio networks. *IEEE Transactions on Parallel and Distributed Systems, 23*(10), 1867–1881.
8. Shin, J., Yang, D., & Kim, C. (2010). A channel rendezvous scheme for cognitive radio networks. *IEEE Communications Letters, 14*(10), 954–956.
9. Stevenson, C. R., Chouinard, G., Lei, Z., Hu, W., Shellhammer, S. J., & Caldwell, W. (2009). IEEE 802.22: The first cognitive radio wireless regional area network standard. *IEEE Communications Magazine, 47*(1), 130–138.
10. Theis, N. C., Thomas, R. W., & DaSilva, L. A. (2011). Rendezvous for cognitive radios. *IEEE Transactions on Mobile Computing, 10*(2), 216–227.
11. Wu, C. -C., & Wu, S. -H. (2013). On bridging the gap between homogeneous and heterogeneous rendezvous schemes for cognitive radios. In *MobiHoc*.
12. Wu, S. -H., Wu, C. -C., Hon, W. -K., & Shin, K. G. (2014). Rendezvous for heterogeneous spectrum-agile devices. In *INFOCOM*.

Chapter 18
Rendezvous Search in a Graph

Abstract The rendezvous search problem in a graph has been widely studied. The problem is defined as follows: two players are initially placed randomly in a space $\mathscr{S}$ which can be represented by discrete points or is continuous. Two players want to meet up, which is the so-called "rendezvous". In a compact space, it is hard to define how exactly two players meet. Therefore, we assume they are said to meet if their distance is no larger than a given value r. This assumption is reasonable because two players can look around and find each other if someone is within the field of vision. r can be considered the detection radius of the player. The goal of rendezvous search is to minimize the time for the players to meet. In this chapter, we first introduce the hardness of rendezvous search in Sect. 18.1, where two types of symmetry are presented. In order to show the intuitive ideas of designing rendezvous search algorithms, we choose rendezvous search along a cycle as the example in Sect. 18.2. The rendezvous search algorithms are presented in Sect. 18.3, and we summarize the chapter in Sect. 18.4.

18.1 Symmetry of Rendezvous Search

There are different settings for the rendezvous search problem, and we introduce two types of symmetry as follows:

(1) *Symmetry of the players*: Similar to the rendezvous process in distributed systems, two players are symmetric if they are indistinguishable; otherwise, they have some special labels or identifiers (IDs).
(2) *Symmetry of the search space*: Similar to the oblivious and non-oblivious setting in distributed systems, the players may see the same information of the search space, which is called symmetry (of the search space).

When two players are symmetric, they have no distinguishable labels and they have to execute the same rendezvous strategy. This corresponds to the algorithm-symmetric setting in distributed systems. When two players are asymmetric, they can agree on different strategies before the problem starts, which corresponds to the algorithm-asymmetric setting. Notice that, we combine the two aspects Alg and ID

© Springer Nature Singapore Pte Ltd. 2017
Z. Gu et al., *Rendezvous in Distributed Systems*,
DOI 10.1007/978-981-10-3680-4_18

in distributed systems since the non-anonymous setting can be used to break symmetry, which means they can be used to design different algorithms, i.e. asymmetric algorithms.

The symmetry of the search space has many different meanings. Different from the distributed systems, the players in a search space can walk along a specific direction such as clockwise or anticlockwise, and face north, south, east or west. Moreover, the search space has both X-axis and Y-axis, but these information can be hard to obtain. Therefore, the symmetry of the search space can be represented as follows according to the available information.

(1) The players can see the coordinates of each point in the space, including the distance along x-axis and y-axis, the origin point (denoted as $(0, 0)$ in the space), and the direction of each movement (corresponds to the forward direction of the axes);
(2) the players can see the x-axis and y-axis with forward direction, but they cannot see the origin point, which means they do not have a common point for rendezvous;
(3) the players can see the x-axis and y-axis, but they do not know the directions of the axises, neither do they know the origin point;
(4) the players cannot see any information about the x-axis and y-axis, but they know the directions, such as clockwise or anticlockwise;
(5) the players can see nothing in the search space, which means they are in a absolutely symmetric space.

Rendezvous in the search space is interesting and these different settings have drawn much attention from the researchers. There is an interesting example proposed in [1]. Suppose there is a straight-line river, over which there is a single bridge. Two players are randomly placed in the river and they want to meet. This is a one-dimension space where only the axis along the river is considered.

For the first situation, the players can see the river and how the river flows. Actually, they can define a direction according to the flow of the river. In addition, they can see the single bridge, which can be regarded as the origin point. This setting is a fully asymmetric one, where the players can design strategies that depend on the initial positions (since they know the origin point and the direction of the axis) and system information.

For the second situation, the bridge is removed and the players do not know their exact positions. But they can use the river and the flow directions to design rendezvous strategies. This setting is harder than the first one.

For the third version, the river stops flowing and the players cannot find the directions of the axes. This makes rendezvous harder than those situations in which they do have the directions.

For the fourth version, suppose the river does not exist in the space, but the players can determine the direction according to some other information, such as how the sun moves.

For the most difficult version, the players have no common direction, no origin point, nor the axis. They have to design algorithms in a fully symmetric situation.

18.2 Rendezvous Search Along a Cycle

In this section, we show a simple example of rendezvous search, and the reader may refer to [1] for more detailed versions of rendezvous search in a graph. We introduce how to design efficient rendezvous search algorithms when two players are randomly placed along a cycle.

As illustrated in Fig. 18.1, two players are placed randomly along a cycle. Suppose the cycle is presented by discrete points, i.e. suppose there are N discrete points along the cycle. Any two adjacent points have distance 1 (a unit length). There are several settings, as discussed above. First, we consider the symmetry of the players.

(1) *Symmetric Players*. Two players are symmetric if they have no distinguishable labels, and they have to run the same algorithm;
(2) *Asymmetric Players*. Two players have different roles in the rendezvous search problem and they can run different strategies.

Similarly, we define the symmetry of the search space (the cycle) as follows:

(1) The cycle has N discrete points and each point is labeled globally, which means the players know the labels of the point beforehand. Suppose the players know the directions of the movement, such as the clockwise direction in Fig. 18.2. In addition, the players know the location of point 0;
(2) The players can see the label of the point that they are at, and they know the clockwise direction. However, they do not know the position of point 0.

Fig. 18.1 An example of rendezvous search along a cycle

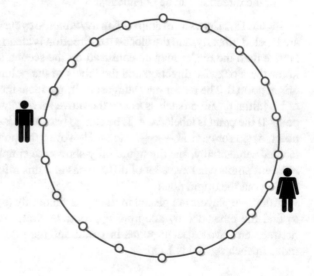

Fig. 18.2 An example of
rendezvous search along an
asymmetric cycle

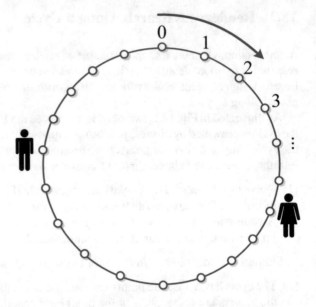

(3) The players can see the label of the point that they are at, but they do not know
     the clockwise direction nor the position of the origin point 0.
(4) The points in the cycle are not labeled and the players can only know the clock-
     wise direction.
(5) The players are in an absolutely symmetric cycle, where no labels of the points,
     or the direction can be obtained.

Figure 18.2 shows an example of an asymmetric cycle where the points in the cycle
are labeled globally, and the clockwise direction is clearly achievable by the players.
Notice that, the reader may be confused by the second situation, where the players
know the clockwise direction and the labels of the points, but they do not know the
origin point 0. The reader may suggest a simple idea: the player can record the label
of his initial position (such as k) and the move forward with step length 1 to the next
point. If the point is labeled $k + 1$, he can go backward $k + 1$ steps to get to the origin
point, or go forward $N - k + 1$ steps. However, the points in the cycle may not be
labeled sequentially, and the figure only shows the simplest version. Therefore, two
adjacent points can have a lot of differences and this information cannot be used for
finding out the origin point.

After two players are placed in the cycle randomly (at any two points), they have
to find the other one by adopting appropriate strategies. We assume the distance
between any two adjacent points is 1 unit and the maximum speed of the players'
move in each second is 1 unit.

18.3 Rendezvous Search Algorithms

We introduce several rendezvous search algorithms in this section, and present how these algorithms can be used for the introduced rendezvous settings.

Algorithm 18.1 Towards Origin Point Algorithm

1: Denote the label of the initial points as k;
2: Compute the direction of smallest length from k to the origin point;
3: Move with the computed direction with maximum speed;

The first algorithm is called the *Toward Origin Point Algorithm*, where the player moves towards the origin point 0 with maximum speed 1. Notice that, there are two directions and the player can compute which direction is better such that the player will spend less time. This algorithm is only feasible when the players have the same labels of the points in the cycle. Notice that, the algorithm can work for two symmetric players. The time complexity to achieve rendezvous is relatively low. The worst situation is: one player has to spend $\lceil \frac{N}{2} \rceil$ seconds to reach the origin point.

The second algorithm works as follows:

(1) One player (denote as A) moves clockwise with unit speed 1;
(2) the other player (denote as B) moves anticlockwise with unit speed 1.

This algorithm assumes that two players are asymmetric and they know the directions along the cycle. Actually, they can rendezvous when they move with opposite directions. Notice that, there are some special situations we should mention.

For example, suppose player A is at point 2 and player B is at point 3 initially. If player A goes clockwisely as in Fig. 18.2, he will get to point 3 in the next second, while player B will get to point 2. In some works, it is assumed that both players can rendezvous within the movement, i.e. they can rendezvous halfway, while some other works assume they can only rendezvous at some point. We do not focus on any special situation in this book. The time used to achieve rendezvous is also small and the worst situation will also use $\frac{N}{2}$ seconds when two users are adjacent, but they move along opposite directions.

When two players do not have a common direction, we propose the third algorithm (known as Wait For Mommy):

(1) One player (denote as A) stays at the initial point;
(2) the other player (denote as B) moves along the cycle in a fixed direction and maximum speed of 1 unit per second.

In the algorithm, player A can be considered as the child who gets lost in a supermarket and he will stay where he is and waits for his mother, while player B represents his mother who searches all places to find her boy. This algorithm is feasible when two players are asymmetric, but we do not know the directions or the labels of the points along the cycle.

All the three algorithms introduced cannot perform well for the fully symmetric situation, where the players are symmetric, and the search space (the cycle) is also symmetric for the users. We present a randomized algorithm for this most difficult situation.

Algorithm 18.2 Random Half Cycle Algorithm

1: Denote the initial point of the player as P;
2: Denote two directions as d_1, d_2;
3: **while** Not Rendezvous **do**
4:    Generate a random value $p \in [0, 1]$;
5:    **if** $p < 0.5$ **then**
6:       Walk with direction d_1 for $\frac{N}{2}$ seconds with maximum speed 1;
7:    **else**
8:       Walk with direction d_2 for $\frac{N}{2}$ seconds with maximum speed 1;
9:    **end if**
10: **end while**

The *Random Half Cycle Algorithm* is presented in Algorithm 18.2. Suppose the initial point of the player is P but he does not know the label of the point. Denote the two directions as d_1, d_2 but he does not know which one means clockwise. For every $\frac{N}{2}$ seconds, the player generates a random value p and it walks along direction d_1 if $p < 0.5$. This means the player does not know which direction is better, and he can only choose a fixed direction for the next $\frac{N}{2}$ seconds with probability 0.5. The player continues the process until rendezvous happens.

This randomized algorithm cannot guarantee rendezvous for certain, but two players running the algorithm can achieve rendezvous with high probability. This is because: when two players choose opposite directions for rendezvous as in Fig. 18.3 for $\frac{N}{2}$ seconds, they can definitely rendezvous (here we omit the details such as whether N is odd or even). However, if the players walk along the same direction, they cannot rendezvous and their relative distance is the same. Therefore, if two players choose opposite directions, they can rendezvous (Fig. 18.4). We show the efficiency and the correctness of the algorithm.

Theorem 18.1 *Two symmetric players running Algorithm 18.2 can achieve rendezvous within $\frac{cN \log N}{2}$ seconds with high probability, where c is a constant.*

Proof In every $\frac{N}{2}$ seconds, the players will make decisions independently. Denote event A as two players who choose the opposite directions every $\frac{N}{2}$ seconds; then:

$$Pr(A) = \frac{1}{2} \qquad (18.1)$$

Actually, we can denote event A_i as two players who choose the opposite directions in the i-th $\frac{N}{2}$ second. It is easy to see that events $A_1, A_2, \ldots, A_i, \ldots$ are independent, and we use A to represent these events. Clearly, if event A happens, two

Fig. 18.3 Two players walk with opposite directions

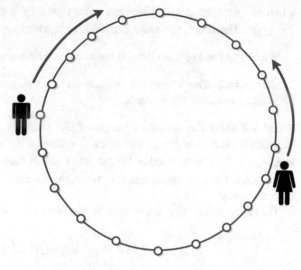

Fig. 18.4 Two players walk with opposite directions

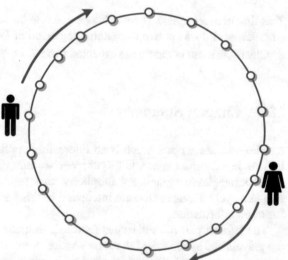

players can rendezvous. Therefore, we suppose event A does not happen for $c \log N$ times where c is a constant, and the probability can be deduced as:

$$Pr\left(\overline{A_1} \bigcap \overline{A_2} \bigcap \ldots \bigcap \overline{A_{c \log N}}\right) = \left(1 - \frac{1}{2}\right)^{c \log N} = \left(\frac{1}{N}\right)^c \qquad (18.2)$$

when $N \to \infty$, rendezvous happens within the i-th $\frac{N}{2}$ second with high probability $1 - (\frac{1}{N})^c$. Therefore, two users can achieve rendezvous with high probability.

We derive the expected time to rendezvous in Theorem 18.2.

Theorem 18.2 *Two symmetric players running Algorithm 18.2 can achieve rendezvous in expected $\frac{3N}{4}$ seconds.*

Proof We show the sketch of the proof. As shown in Theorem 18.1, rendezvous happens if event A happens. For every $\frac{N}{2}$ seconds, the expected time to rendezvous is $E(A) = \frac{N}{4}$ seconds based on the fact that event A must happen. This can be verified easily since the average situation is when the distance between two players are $\frac{N}{2}$ units far away.

We suppose the expected time to rendezvous is T, and we formulate it as:

$$T = Pr(A) * E(A) + Pr(\overline{A}) * \left(T + \frac{N}{2} \right) \tag{18.3}$$

This first item represents event A happens while the second item computes the time when event A does not happen within the period of $\frac{N}{2}$ time slots.

Clearly, we can compute the expected time $T = \frac{3N}{4}$, and so the theorem holds.

18.4   Chapter Summary

Rendezvous search in a graph is an interesting application, where two players try to meet in a compact space. In this chapter, we introduce how to design rendezvous search strategies in a graph. For simplicity, we consider the rendezvous search problem in a cycle. Readers who are interested in this particular subject can refer to [1] for more information.

The hardest part of rendezvous search is to break symmetry, both symmetry of the players and symmetry of the search space. Similar to rendezvous in a distributed system, the players' IDs can be used to break symmetry among the players and the labels of the search space can be used to break symmetry of the space. We list different settings to point out the hardness of rendezvous search in this chapter.

We propose two different rendezvous search strategies for rendezvous search along a cycle. The first algorithm is called Towards Origin Point Algorithm where the origin point is known to both players and they can walk towards the origin point for rendezvous. The second one is called Random Half Cycle Algorithm where the players have no outside (or global) information and they have to walk along the cycle in different directions based on predefined probability. This algorithm can make two players find each other in a short expected time.

There are many interesting variants of the rendezvous search problem, such as to minimize the time to find each other, and two players trying to fulfill a rendezvous search game, where one player wants to be meet quickly but the other one does not want to be found. Such applications or problems can utilize the intuitive idea of basic rendezvous in distributed systems, and we hope the ideas and techniques introduced here can be to solve more such related problems.

Reference

1. Alpern, S., & Gal, S. (2003). *The Theory of Search Games and Rendezvous*. Berlin: Springer.

Chapter 19
Neighbor Discovery in Wireless Sensor Networks

Abstract Wireless sensor networks (WSNs) are widely used in many applications such as air pollution monitoring, natural disaster prevention, health care monitoring, etc. The sensor nodes are deployed in the monitored area and they form a wireless network through communication with nearby sensors. In this chapter, we introduce the fundamental process in constructing a wireless sensor network, which is called **neighbor discovery**, where the sensors can find nearby neighboring sensors when their distance is within a threshold. There are two main reasons for studying the neighbor discovery problem. First of all, the deployed sensors as a configuration may vary dynamically according to different reasons. For example, new sensors may be added and old ones removed; and some types of sensors have the ability to move around inside the area. Therefore, a sensor may need to find nearby neighboring sensors when they move into its communication range or some new sensors are deployed in the vicinity. The second reason is the limited power supply. In most situations, sensors are powered by battery and their energy is very limited. For example, suppose a sensor is powered by a 1200 mAh battery, and the processor consumes 2 mA under full power and the radio consumes 20 mA when it is turned on. If the sensor keeps the radio on and computing never stops, the lifetime of the sensor is a little more than two days. Therefore, sensors need some special method to save energy in order to extend their lifetime. A trivial way is to add sleeping mode, where a sensor keeps silent in a sleep state for most of the time, wakes up for work for only a small fraction of the time. Practically, in sleep mode, the processor's power consumption drops to $2\,\mu A$ and the radio power drops to $1\,\mu A$, and thus lifetime can be extended significantly. If the sensor wakes up for only 1% of the time and sleeps for 99% of the time, the estimated lifetime would be half a year. If the wake-up portion is 0.1% of the time, the lifetime can be over five years. Therefore, we assume sensors are silent for most of the time, and they wake up mainly for data collection and communication. The neighbor discovery problem is to design the schedules of the sensors' sleep mode and wake mode, such that two nearby sensors can be in wake mode at the same time to find each other, which is a kind of rendezvous problem. In this chapter, we first propose a motivational example in Sect. 19.1, and formulate the problem in Sect. 19.2. We introduce the trivial brute force algorithm for the neighbor discovery problem in Sect. 19.3 and another two algorithms: relaxed difference set based algorithm and co-prime algorithm in Sects. 19.4 and 19.5 respectively. Finally, we summarize the chapter in Sect. 19.6.

© Springer Nature Singapore Pte Ltd. 2017   243
Z. Gu et al., *Rendezvous in Distributed Systems*,
DOI 10.1007/978-981-10-3680-4_19

19.1 Motivational Example

In sensor networks, the sensors may turn on or off their radio periodically, randomly or according to some pre-defined rules. In Fig. 19.1, three sensor nodes have different wake-up schedules, where the top lines of each node mean the radio is on while the bottom lines mean the radio is off. The sensor nodes use different wake-up schedules and two neighboring users can find each other if they are close enough and their radios are both turned on at the same time. Suppose all three nodes are near to each other, and they can discover each other if they turn on the radio at the same time. As shown in the figure, node a and node b may find each other when they both turn on the radios, but node a and node c cannot find each other since their wake-up times do not intersect. We call the latter uncoordinated schedules which could be the result of improper initialization, crash at unpredicted times, etc.

Some MAC protocol research proposed a sampling method where a sensor node would sample radio activities during its sleeping times, to try to learn their wake-up schedules of nearby nodes. If the neighboring nodes' schedules can be predicted, the node can add some additional wake-up times to its own schedule to achieve coordination with the other nodes. For example, node a in can learn the other two nodes' schedules during its sleeping times, and it will revise its wake-up strategy such that it can discover its neighbors (see Fig. 19.1).

Although the learning method can help solve the neighbor discovery problem, it increases the ratio of wake-up time, which leads to shorter lifetime. Therefore, in this chapter, we study how to design efficient algorithms such that the sensor nodes can finally discover each other with a coordinated wake-up schedule, while keeping a low wake-up ratio.

In designing neighbor discovery algorithms for sensor networks, we face the following challenges:

(1) Symmetric schedule: the sensors nodes should adopt the same wake-up schedule since they are anonymous and placed in a distributed fashion in the monitoring area;
(2) asynchronous time clock: each sensor node may have a local time clock, and they will execute according to its own time clock.

In many practical applications, a large number of sensor nodes are deployed in the monitoring area and it is impossible to design a special wake-up schedule for each

Fig. 19.1 An example

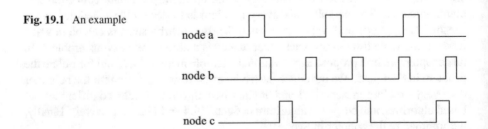

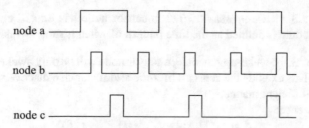

Fig. 19.2 An example of learning the neighboring nodes' wake-up schedule

sensor node. Therefore, a symmetric schedule for all sensors is preferred. Different sensor nodes may be deployed at different times, their local clocks are different and they will execute the neighbor discovery algorithm according to its own clock. The designed algorithms should take all this into account.

19.2 Problem Definition

We define the terms and variables used to formally describe the neighbor discovery problem for wireless sensor networks.

Definition 19.1 (*Time slot*) we divide time into slots of equal length, where two neighboring nodes can discover each other during one unit time slot if they both have their radios turned on.

Definition 19.2 (*Slot aligned network*) suppose there exists a global clock where the time is divided into slots of equal length. For any node in the network, the start time of its own clock must be the start time of any slot in the global clock.

For example, in Fig. 19.3, the start time of node a's local clock meets the start time of the global time slot 3, while node b's start slot meets the start time of the global time slot 6. If some nodes are not slot aligned, choose a nearby time slot to continue. This slot aligned idea is also used in traditional rendezvous.

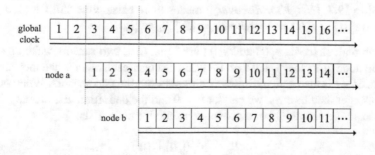

Fig. 19.3 An example of slot aligned network

Definition 19.3 (*Wake-up schedule*) each sensor node has limited energy and the wake-up schedule is defined as the time pattern of when it should wake up or sleep.

By designing the wake-up schedule, a sensor node will be only awake for a portion of the time. This extends the sensor's lifetime and it can also discover its neighbors in an energy-efficient manner.

For a sequence:

$$S = \{s(1), s(2), \ldots, s(t), \ldots, s(T)\}$$

where $s(t)$ means the state of the sensor at time t, we formally define the wake-up schedule as such:

Definition 19.4 (*The neighbor discovery (wake-up) schedule*) of the sensor node u_a is defined as a sequence $S_a = \{s_a(1), s_a(2), \ldots, s_a(t), \ldots, s_a(T)\}$ where T is the length and each element $s_a(t)$ is:

$$s_a(t) = \begin{cases} 0 \text{ if sensor } u_a \text{ sleeps in slot } t \\ 1 \text{ if sensor } u_a \text{ wakes up in slot } t \end{cases}$$

Definition 19.5 (*Duty cycle*) d_a of the neighbor discover schedule S_a is defined as:

$$d_a = \frac{|\{1 \leq t \leq T : s_a(t) = 1\}|}{T}$$

We define the asynchronous situations of time slots as:

Definition 19.6 (*Clock drift*) means every two neighboring sensor nodes may have random clock drifts. Suppose node u_a is k time slots earlier than node u_b, we can rotate the neighbor discovery schedule by k slots as:

$$rotate(S_a, k) = \{s_a^k(t) | s_a^k(t) = s_a((t + k - 1) \mod T + 1), t \in [1, T]\}$$

Suppose two sensor nodes u_a and u_b have corresponding neighbor discover schedules S_a and S_b, and user u_a is k time slots earlier than node u_b; we define the neighbor discovery problem between them as:

Definition 19.7 (*Neighbor discovery*) means there exists time slot $t \in [1, T_a * T_b]$ such that $s_a^k(t) = s_b(t) = 1$ where $T_a = |S_a|, T_b = |S_b|$.

When both elements $s_a^k(t)$ and $s_b(t)$ are equal to 1, two sensors wake up at time slot t (in node u_b's clock), and then they can discover each other when they are close enough. Slot t is also called discovery slot between the two sensors. When node u_a is k slots later than user u_b, we can let $k < 0$ and the definition also works.

For example, the neighbor discover schedule for node u_a is:

$$S_a = \{0, 0, 0, 1, 0\}$$

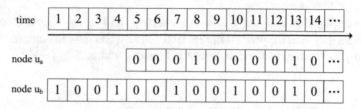

Fig. 19.4 An example of neighbor discovery between two synchronous nodes

time	1	2	3	4	5	6	7	8	9	10	11	12	13	14	⋯
node u_a				0	0	0	1	0	0	0	0	1	0		⋯
node u_b	1	0	0	1	0	0	1	0	0	1	0	0	1	0	⋯

Fig. 19.5 An example of neighbor discovery between two asynchronous nodes

while the neighbor discover schedule for node u_b is:

$$S_a = \{1, 0, 0, 1, 0, 0\}$$

As shown in Fig. 19.4, if there is no clock drift between two nodes, they can find each other in time slot 4, and Fig. 19.5 shows the example when node u_a is 4 time slots earlier than node u_b. Clearly, two nodes can also discover each other in time slot 13 (and it is time slot 9 in node u_a's clock). In these two examples, two sensor nodes adopt different wake-up schedules which is not that practical. Symmetric wake-up schedule is preferred by two and definitely by more nodes.

19.3   Brute Force Algorithm

Since the sensor nodes may start the neighbor discovery algorithm at any time, if the sensor nodes keep awake all the time, two neighbors can clearly find each other. However, the duty cycle then of the sensors is 100%, which is inefficient. An easy improvement is to make the sensor nodes wake up for k time slots, where $k \geq \lceil \frac{N+1}{2} \rceil$. We describe the method as follows:

(1) Time is divided into rounds and each round contains N time slots;
(2) in each round, the sensor node keeps awake in the first $k \geq \lceil \frac{N+1}{2} \rceil$ time slots and asleep for the other time slots.

Notice that, when $k = \lceil \frac{N+1}{2} \rceil$, two different sensors should have at least one common slot when they are all awake, and this is called the "51%" solution, since the node keeps awake for more than half of the time in each round.

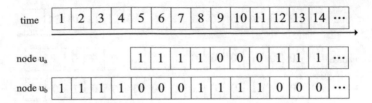

Fig. 19.6 An example of brute force algorithm

For example, if $N = 7$ and $k = 4$, two nodes u_a and u_b can find each other no matter when they start the process in Fig. 19.6. Clearly, it is easy to compute the time complexity to discover each other, which is bounded within $T \leq \lceil \frac{N+1}{2} \rceil$.

19.4 Relax Difference Set Based Algorithm

Another method is to design a neighbor discovery schedule on the basis of relaxed difference set (RDS). Recall the definition of RDS:

Definition 19.8 A set $D = \{a_1, a_2, \cdots, a_k\} \subseteq Z_n$ (the set of all nonnegative integers less than n) is called a Relaxed Difference Set (RDS) if for every $d \neq 0 \pmod{n}$, there exists at least one ordered pair (a_i, a_j) such that $a_i - a_j \equiv d \pmod{n}$, where $a_i, a_j \in D$.

For any n time slots, we can design a fixed relaxed difference set under Z_n and the sensors nodes wake up according to the difference set can find each other efficiently. We describe the algorithm in Algorithm 19.1.

Algorithm 19.1 Relaxed Difference Set Based Algorithm

1: Choose a positive integer $n > 0$;
2: Construct a relaxed difference set $D = \{a_1, a_2, \cdots, a_k\} \subseteq Z_n$;
3: Denote time slot $t := 0$;
4: Denote the constructed neighbor discovery schedule as $S := \{s(0), s(1), \ldots, s(t), \ldots\}$;
5: **while** Not discover the other node **do**
6:     $t' := t \bmod T$;
7:     $index := t' \bmod n$;
8:     **if** $index \in D$ **then**
9:         $s(t) := 1$;
10:    **else**
11:        $s(t) := 0$;
12:    **end if**
13: **end while**

As shown in Algorithm 19.1, we divide time into rounds of n time slots and construct a relaxed difference set under Z_n. Then, we transform the relaxed difference set D into a schedule of length n, as follows:

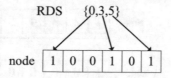

Fig. 19.7 An example of transforming an RDS to a neighbor discover schedule

$$s(t) = \begin{cases} 1 \text{ if } t \in D \\ 0 \text{ otherwise} \end{cases}$$

where $t \in [0, n)$. The sensor nodes repeat the sequence until they find the others. We illustrate the transformation in Fig. 19.7.

According to the definition of RDS, when one node is δ time slots later or earlier than another neighboring node, they can both find each other. We analyze the time complexity for discovering each other and the duty cycle of the designed schedule.

Obviously, no matter when the sensor nodes start the algorithm, they can find each other within n time slots, since the property of RDS guarantees that at least one common time slot exists such that they are awake during n consecutive time slots.

As described in Chap. 8, the size of the relaxed difference set can be as small as $\Omega(\sqrt{n})$ and the duty cycle of the constructed schedule is:

$$d = \frac{|RDS|}{n} = \frac{\Omega(\sqrt{n})}{n}$$

Actually, Chap. 8 shows the method of constructing an RDS with size $\sqrt{3n}$ and thus the duty cycle can be small.

Therefore, there is a tradeoff between the time complexity to discover each other and the duty cycle of the schedule. If we want to reduce the time of discovery, n should be small. If we want to reduce the duty cycle, we should increase n. Considering both, we can choose an appropriate n value in practice to implement the wake-up schedule.

19.5 Co-Prime Algorithm

Co-prime algorithm is based on the Chinese Remainder Theorem as discussed in Chap. 9. If two nodes use two different prime numbers to design their neighbor discovery schedules, they can find each other with very short duty cycle, if the chosen primes are different. This type of algorithms is as follows:

- For sensor node u_a, choose a random prime number p_a;
- time is divided into rounds where each round contains p_a time slots;

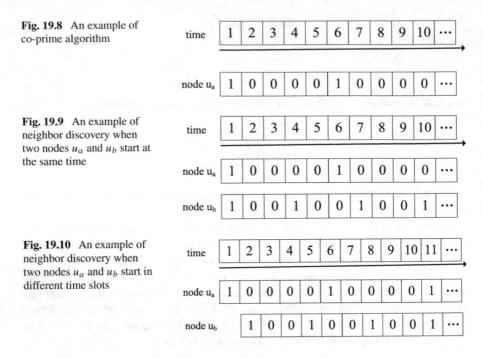

Fig. 19.8 An example of co-prime algorithm

Fig. 19.9 An example of neighbor discovery when two nodes u_a and u_b start at the same time

Fig. 19.10 An example of neighbor discovery when two nodes u_a and u_b start in different time slots

- the sensor node u_a wakes up in the first time slot of each round, and keeps asleep for the other $p_a - 1$ time slots.

For example, as illustrated in Fig. 19.8, if the node u_a chooses prime number $p_a = 5$, the constructed schedule is shown in the figure.

Considering any two neighboring nodes u_a and u_b, denote their chosen prime numbers as p_a and p_b respectively. If $p_a \neq p_b$, they can definitely find each other within $p_a * p_b$ time slots.

For example, node u_a chooses $p_a = 5$ and node u_b chooses $p_b = 3$; they can find each other in the first time slot if they start at the same time, as shown in Fig. 19.9, or they can discover the other in time slot 11 if node u_b is one time slot later than node u_a as depicted in Fig. 19.10 (the time complexity is actually 10 since node u_b is later).

We analyze the time complexity of neighbor discovery and the length of the duty cycle. Suppose node u_a is δ time slots earlier than node u_b. From node u_b's clock, node u_b will wake up in time slot $1 + p_b * l_b$ where p_b is the chosen prime number and l_b can be any positive integer. (Notice that, we assume time starts with slot 1, and sometimes we can also assume it starts with slot 0.) We rewrite the equation as:

$$t \equiv 1 \mod p_b$$

where node u_b wakes up in time slot t.

We then analyze when node u_a would wake up in node u_b's clock. When:

$$(t + k) \mod p_a \equiv 1$$

Fig. 19.11 An example of neighbor discovery does not happen if two sensor nodes choose the same prime number

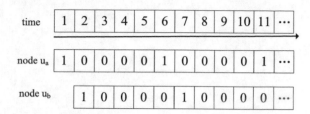

where p_a is the chosen prime number, node u_a will be awake. We rewrite it as:

$$t \equiv 1 - k \mod p_a$$

where node u_a wakes up in time slot t.

Therefore, we need to compute such value t by combining both equations, and there exists $t \in [1, p_a * p_b]$ satisfying both equations from the Chinese Remainder Theorem. Therefore, the time complexity is bounded within $p_a * p_b$.

Obviously, the duty cycle of node u_a is $d = \frac{1}{p_a}$ which can be relatively short. But larger p_a value may lead to higher time complexity of neighbor discovery. Therefore, choosing the appropriate p_a value is important.

Notice that, if two sensor nodes choose the same prime number, they may not find each other if they do not start at the same time. As shown in Fig. 19.11, both nodes choose prime 5 and node u_b is 1 time slot later than node u_a, and they can never find each other. Therefore, the chosen prime numbers should be different, but it is hard to guarantee that the chosen primes are always different in practical wireless sensor networks.

19.6   Chapter Summary

In this chapter, we study neighbor discovery problem in wireless sensor networks. Since the sensors have only limited battery power, if they keep awake all the time, the sensors will go down soon. Therefore, we need to design wake-up schedules for the sensors to save energy and extend their lifetime. However, if different sensors wake up according to different schedules, two neighboring sensors may not discover each other if one of them happens to be asleep. The goal of neighbor discovery problem is to design efficient wake-up schedules for the sensors such that they can find each other in a short time, while having a low duty cycle of being awake.

We introduce three types of rendezvous algorithms. The first one is to be awake for most of the time (larger than 50%) and the sensors can find each other for sure. This type of algorithm is called the brute force algorithm. The second one is relaxed difference set based algorithm, where we use relaxed difference set to design wake-up schedules and the duty cycle can be reduced to $O(\frac{1}{N})$. The third one is co-prime algorithm where the sensors choose prime numbers to generate their wake-up schedules and it guarantees quick discovery if two sensors' chosen prime numbers are different.

Part V
Conclusions and Future Works

Chapter 20
Conclusions and Future Works

In this book, we introduce the fundamental process in constructing a distributed system, which is referred to as rendezvous. A distributed system is composed of multiple autonomous entities that can make decisions locally. Through appropriate cooperation among these distributed entities, the system can be utilized to carry out some global computational tasks. In order for the entities to cooperate, communication links have to be made use of dynamically and effectively through their connected external ports; entities may fail to connect to the others if they make a wrong choice of the external ports. To be able to access the correct connected pair of external ports by the two entities involved in a communication act is essential in all cooperation, and the aim of the rendezvous algorithms we study is to achieve that efficiently.

20.1 Conclusions

In studying the rendezvous problem in distributed systems, there are five important components that should be taken into consideration.

First, the distributed entities could choose to run *symmetric* or *asymmetric* algorithms, where the entities utilize the same algorithm for rendezvous attempt if the choice is symmetric, or they would run different algorithms if it is asymmetric. Obviously, designing symmetric algorithms for the distributed entities would be much more challenging, which however is much more practical for real distributed system implementations. In this book, we present the differences of designing these two types of rendezvous algorithms, and many elegant methods are presented.

Second, timing plays an important role in a distributed system. For simplicity, we assume time is divided into slots of equal length and the distributed entities can perform their local computations in each time slot. Considering different applications, we study two different types of distributed systems: *synchronous* systems where all entities start the algorithm at the same time, and *asynchronous* systems where the entities could start the algorithm in different time slots. Obviously, designing efficient rendezvous algorithms for synchronous systems is much easier than those for

© Springer Nature Singapore Pte Ltd. 2017
Z. Gu et al., *Rendezvous in Distributed Systems*,
DOI 10.1007/978-981-10-3680-4_20

asynchronous systems. However, most practical distributed systems are asynchronous and the entities have to perform their own tasks in an asynchronous manner. In this book, we present simple algorithms for synchronous entities, and introduce relatively more complicated rendezvous algorithms for asynchronous entities.

Third, the entities in a distributed system are not able to utilize all the external ports for communication, this is because some ports may be occupied by other services or other types of communication. We say the external ports are *available* if they are not occupied by other services or communication. In this book, we study the *symmetric port scenario* where all entities in a distributed system have the same set of available external ports, and the *asymmetric port scenario* where different entities may have different sets of available ports. Obviously, the symmetric port scenario is a special case of the asymmetric port scenario, and designing rendezvous algorithms for the latter could be much more difficult than for the former. However, in the so called oblivious blind rendezvous problem (in Part III), the symmetric port scenario could present as much difficulty as the asymmetric port scenario. We present distributed rendezvous algorithms, including deterministic and randomized ones for both the symmetric port scenario and the asymmetric scenario.

Fourth, breaking symmetry is an essential act in distributed systems. We adopt a commonly used assumption to break symmetry where each entity in a distributed system can be assigned a distinguishable identifier (ID). There are two types of distributed systems: *anonymous* systems assume all entities in the system are indistinguishable and they cannot be separated by deterministic methods; *non-anonymous* systems on the other hand assumes each entity in the system has a unique ID which can be used to break symmetry. Designing rendezvous algorithms for non-anonymous systems is much easier than for anonymous systems, since the users' IDs can be used to design entity-specific algorithms. In this book, we present elegant results for both anonymous and non-anonymous systems, but only randomized algorithms are applicable in handling the oblivious blind rendezvous situation in anonymous systems. These randomized algorithms also have good performance on average.

The last component is port labeling, which defines the blind rendezvous problem in Part II and the oblivious blind rendezvous problem in Part III. In the blind rendezvous problem, all entities in the distributed system are assumed to have the same labels of the external ports and any pair of ports with the same label are connected. This situation is called *non-oblivious port labeling*. However, in general distributed systems, the entities inside the system are not able to obtain any global information and so they cannot see the global labels of the external ports beforehand. We call this *oblivious port labeling* where the entities in a distributed system are free to label their own ports according to local rules and they do not know the connection patterns of any pair of the external ports. This blindness of port labeling makes rendezvous much harder to tackle and we call it the oblivious blind rendezvous problem which we study in Part III. In this book, we introduce a series of distributed algorithms for the blind rendezvous problem, and present a few extant results for the oblivious blind rendezvous problem.

To sum up, we study rendezvous in the distributed systems in three parts (Parts II, III, and IV):

In Part II, we formulate the blind rendezvous problem in distributed systems, where the entities in the system cannot see the others' information and they have to execute the algorithms in a "blind" way. In this part, we assume all entities see the global labels of the external ports in advance, and they can run algorithms that build on those global labels. Since rendezvous mainly happens between any two neighboring entities, we provide intuitive ideas, common used techniques, and efficient algorithms for two entities in Chaps. 6–9, and we present an extension for multiple entities in a multihop distributed system in Chap. 10.

The intuitive ideas of handling blind rendezvous for two entities come from the following aspects:

(1) In designing asymmetric algorithms, one entity waits on a fixed port for a sufficiently long time while the other entity accesses the ports sequentially. This method could reduce the time complexity of rendezvous and it is from the simple idea of wait-for-mummy: when a child gets separated with his mother in the supermarket, the child should stay put and wait for his mother to come to meet him. This trivial idea helps the design of efficient asymmetric algorithms for the distributed systems.
(2) In designing symmetric algorithms for anonymous entities, the global labels of the external ports can be utilized to design such algorithms. The idea is to convert the designed rendezvous sequence to a disjoint relaxed difference set (DRDS), and the construction of the DRDS corresponds to the rendezvous sequence that guarantees rendezvous with high efficiency.
(3) In designing symmetric algorithms for non-anonymous entities, the entities' identifiers (IDs) could be utilized to design the rendezvous sequence, by combining the global labels of these external ports. Due to the uniqueness of the entities' IDs, the rendezvous algorithms could be much more efficient than those for anonymous entities.

The techniques introduced in designing rendezvous algorithms include:

(1) Channel hopping, which is originally defined for cognitive radio networks, can also be used for general distributed systems. Since the rendezvous algorithms have to be suitable for asynchronous entities, they have to generate a sequence with fixed length and repeat the sequence to access the ports. Normally, the sequence is generated by some mathematical calculations which can be considered as hopping to the next channel (or port in the distributed system) by adding some hopping step. This technique is commonly utilized in designing various types of rendezvous algorithms.
(2) ID Scaling, which converts the entity's ID to another unique string, is often used in designing symmetric algorithms for non-anonymous entities. The converted strings of different entities must have different bits and this kind of information can be used as different hopping steps in generating the rendezvous sequences.

This book introduces many kinds of rendezvous algorithms. For asymmetric algorithms, we shed new light on designing efficient algorithms that have very low time complexity. For symmetric algorithms, we present global sequence (GS) based rendezvous algorithms for anonymous entities and local sequence (LS) based rendezvous algorithms for non-anonymous entities. The GS based algorithms can guarantee rendezvous within $O(N^2)$ time slots, which matches the known derived lower bound of such kind of algorithm. The LS based algorithms guarantee rendezvous within $O((\max\{|C_A|, |C_B|\})^2)$ time slots, where C_A, C_B represent the available port sets for the two entities respectively. This result is much better than the GS based algorithms especially when the available ports only account for a small fraction of all the ports.

In Part III, we formulate the oblivious blind rendezvous problem in distributed systems, where the entities in the system cannot see the global labels of the external ports beforehand, and they have to label the external ports locally. This part tackles a harder but more general rendezvous setting. Actually, when the entities have no information about the external ports, the method of generating a rendezvous sequence on the basis of the ports' labels cannot work. For some worst-case scenarios, there simply does not exist any deterministic distributed algorithm and the rendezvous cannot be guaranteed within a finite time. In this part, we focus on the oblivious blind rendezvous problem between two entities, in Chaps. 12–15, and we discuss the extension to multiple entities in a multihop distributed system in a simple way in Chap. 16.

The intuitive ideas behind handling oblivious blind rendezvous for two entities come from the following:

(1) In designing asymmetric algorithms, we adopt the strategy that one entity should wait on a fixed port for a sufficiently long time, while the other entity accesses the available ports sequentially. The hardness comes from tuning the relevant parameters such that the waiting time is long enough to cover the possible accessing of the ports by the other entity; but long waiting time could lead to long rendezvous time, since the entity does not know which port is the correct port to wait for.

(2) In designing symmetric algorithms for non-anonymous entities, the key point is to break symmetry between the entities, and their IDs play the vital role. In addition, some mathematical principles, such as co-prime properties, should be taken into consideration. Actually, if two entities repeat a sequence of different lengths $p_a \neq p_b$ and p_a, p_b are two prime numbers, they can definitely rendezvous. This observation is also a novel idea in designing deterministic distributed rendezvous algorithms.

(3) In designing symmetric algorithms for anonymous entities, it is impossible to design deterministic ones, which is also shown in the book. Therefore, tricky randomized methods are employed to design rendezvous algorithms with expected low time complexity. The intuitive idea is to sample different prime numbers or different randomized values that are coupled in the rendezvous sequences.

The techniques introduced in designing oblivious blind rendezvous algorithms include:

(1) ID Scaling, which is also an important technique in blind rendezvous algorithms, converts an entity's ID to another unique string, and the corresponding rendezvous sequence can be created on the basis of the string.
(2) Randomized sampling, which generates different random values or random prime numbers, is commonly adopted in breaking symmetry between two anonymous entities. The difference of sampled randomized numbers could help design rendezvous sequences and guarantee bounded rendezvous. However, such sampling can only guarantee achieving rendezvous with high probability, but it cannot assure rendezvous for any situation.

Different kinds of oblivious blind rendezvous algorithms are presented in Part III. For asymmetric algorithms, we extend the design of blind rendezvous algorithms while keeping the increased time complexity at an acceptable level; about $O(N^2)$ time slots are sufficient to achieve rendezvous. For symmetric algorithms between two non-anonymous entities, building on the converted bits of the entities' IDs, we construct rendezvous sequences with different hopping steps and the time complexity is about $O((\max\{k_a, k_b\})^2)$ time slots for most situations, where k_a, k_b represent the two entities' number of available ports respectively. In addition, we show the method of deriving lower bounds for such rendezvous algorithms and the proposed algorithms have acceptable performance compared with the refined bound $\Omega((k_a - k_g) \cdot (k_b - k_g))$ where k_g represents the number of common available ports. For symmetric algorithms between two anonymous entities, we design several randomized algorithms and the efficiency is also guaranteed through careful calculations.

In Part IV, we discuss three practical distributed systems where rendezvous process plays a fundamental role. The first one is a heterogeneous cognitive radio network (HCRN), which is a special type of cognitive radio network. In the HCRN, each entity is equipped with a cognitive radio that can sense a set of continuous licensed channels. We modify the blind rendezvous algorithms such that the time complexity could be much smaller than that of the GS based algorithms. Then, we introduce rendezvous search in a graph where the entities should meet at some discrete locations in the graph. For simplicity, we present several rendezvous algorithms for the rendezvous search along a cycle and we compare the rendezvous problem in different situations with the studied rendezvous settings in Part II and Part III. The last application we consider is the neighbor discovery problem in wireless sensor networks, where the sensors should have low duty-cycle schedule to prolong their lifetime. Though it is not related to the rendezvous problem directly, we can regard the design of wake-up schedules as the design of rendezvous sequences. Different types of rendezvous based algorithms are also introduced for designing the schedule such that two nearby sensors can discover each other, while keeping to a relatively low duty-cycle.

20.2   Future Works

This book introduces the rendezvous problem in the distributed systems, which also covers rendezvous theory, efficient distributed algorithms, and some representative distributed applications that use rendezvous. Through the chapters, the readers can fully understand what role the rendezvous process plays in constructing a distributed system, what the rendezvous problem tries to solve from both theoretical and practical views, how rendezvous algorithms work for the autonomous entities in the system, to what extent rendezvous efficiency could be achieved, and how the rendezvous process can be applied in the other application domains. In order to broaden the scope of the rendezvous theory, the algorithms, and the applications, we suggest several possible future directions and the readers who are interested in this subject may follow the points given to pursue further.

First of all, closing the gap between the lower bounds on the maximum time to rendezvous in worst-case situations and the upper bounds by the presented algorithms will likely be a long term project. For example, as shown in Sect. 8.6, the lower bound of the global sequence based rendezvous algorithms could be as low as (about) N^2 time slots through theoretical analysis. However, the state-of-the-art result can only guarantee rendezvous within (about) $3N^2$ time slots for the worst situation. The gap between these two bounds cannot be closed easily at least for now. We should try to find tighter lower bounds or design more efficient algorithms. In Table 8.2, we compute the relationship between the value n and the maximum cardinality of the disjoint relaxed difference set under Z_n when n is not large, which corresponds to the lower bound of the global sequence based algorithm; the table implies that it is hard to derive tighter theoretical lower bound, but one meaningful direction would be to design shorter global sequences that can be adopted in rendezvous.

Considering the oblivious blind rendezvous between two non-anonymous users as another example, the derived lower bound by the adversary assignment graph in Sect. 13.3 requires at least $\Omega((k_a - k_g) \cdot (k_b - k_g))$ time slots to rendezvous, where k_a, k_b represent the number of available ports for the two users respectively and k_g represents the number of common available ports between them. However, when the users have no global labels of the external ports, i.e. oblivious port labeling, the extant results can only guarantee rendezvous within $O((\max\{k_a, k_b\})^2)$ time slots [5] or within $O(k_a k_b)$ time slots under certain assumptions [4]. The gap between the lower bound and the upper bounds is large. The direction of deriving tighter lower bounds and designing better algorithms is an important one.

Besides the theoretical bounds for the rendezvous problem, another major area is the handling of dynamic occupancy of the external ports. For example, the following two entities have different sets of available ports:

$$\begin{cases} C_a = \{1, 3, 5, 7, 9\} \\ C_b = \{2, 4, 6, 8\} \end{cases}$$

Suppose they utilize a simple rendezvous algorithm where each entity access the available port sequentially, the users can never rendezvous since they have no common available port. However, the occupancy of the external ports by other services may vary over time and the set of the available ports could also change. Suppose port 9 becomes available for the second entity in time slot 3 and they can achieve rendezvous in time slot 5. This dynamic change can help achieve rendezvous, but it can also reduce the chance of communication when some common available ports are occupied temporarily. How to accommodate such dynamic changes of the ports' occupancy is an interesting and useful problem. We believe this could be a promising direction in the future, not only due to the lack of a corresponding rendezvous theory, but also the phenomenon occurs in many practical distributed systems. One possible method is to handle the changes of the ports' states by randomized sampling, while the other strategy would be generating the change patterns through machine learning.

Other than theory and algorithms in further rendezvous research, many interesting applications are also worth further exploring.

The first related application would be the classic telephone coordination problem which is introduced in Sect. 1, and the best known result, the Anderson-Weber strategy (AW), is proposed in [3]. However, this work only proposes a good randomized strategy that has better performance than random selection, i.e. $0.829n$ time slots are needed in expectation compared with n time slots. It is still an open problem whether a better algorithm exists with even shorter expected time.

Another application is the multi-radio cognitive radio network, where each user in the network is equipped with multiple cognitive radios that can switch to multi-channel mode for rendezvous in one time slot. This new network architecture can help reduce rendezvous time and enable more efficient rendezvous protocols. However, there are very few related works [6–9] and extant research lacks comprehensive theoretical foundations for this relatively new type of networks.

The last application we discuss is rendezvous search game [1, 2]. In Chap. 18, we introduce rendezvous search in a graph where two players aim to meet at a discrete location of the graph as quickly as possible. However, the rendezvous search game turns the problem into a complicated one. Similar to rendezvous search, two players are involved but they have different goals: one player tries to find the other player as quickly as possible while the other player tries to "hide" and avoid being found. The different goals of the players make the problem interesting, but harder to solve. This is also why it is called *a game*. Some works have studied the problem but more general solutions remain missing.

The rendezvous process is indeed vital in constructing a basic distributed system. We talk about the theory of rendezvous, present distributed algorithms that have high efficiency, and discuss some interesting applications where rendezvous can be adopted. This field is still at a tender age from the perspective that it is still far from widely applied, and so to induce deeper understanding in a wider audience is important for its further development. This book represents a small step of that mission, and we hope the readers can acquire a good understanding of the rendezvous problem and its solutions after reading the book.

References

1. Alpern, S., & Lim, W. S. (1998). The symmetric rendezvous-evasion game. *SIAM Journal of Control and Optimization, 36*(3), 948–959.
2. Alpern, S., & Gal, S. (2003). *The theory of search games and rendezvous.* Berlin: Springer.
3. Anderson, E. J., & Weber, R. R. (1990). The rendezvous problem on discrete locations. *Journal of Applied Probability, 28*, 839–851.
4. Chen, L., Bian, K., Chen, L., Liu, C., Park, J.-M. J., & Li, X. (2014). A Group-theoretic framework for rendezvous in heterogeneous cognitive radio networks. In *MobiHoc.*
5. Gu, Z., Hua, Q.-S., & Dai, W. (2014). Fully distributed algorithms for blind rendezvous in cognitive radio networks. In *MOBIHOC.*
6. Li, G., Gu, Z., Lin, X., Pu, H., & Hua, Q-S. (2014). Deterministic distributed rendezvous algorithms for multi-radio cognitive radio networks. In *MSWiM.*
7. Paul, R., Jembre, Y. Z., & Choi, Y.-J. (2014). Multi-interface rendezvous in self-organizing cognitive radio networks. In *DySPAN.*
8. Yu, L., Liu, H., Leung, Y.-W., Chun, X., & Lin, Z. (2013). Multiple radios for effective rendezvous in cognitive radio networks. In *ICC.*
9. Zhang, J., & Zhang, Z. (2011). Initial link establishment in cognitive radio networks without common control channel. In *WCNC.*

Printed in the United States
By Bookmasters